# Lecture Notes in Energy

Volume 42

Lecture Notes in Energy (LNE) is a series that reports on new developments in the study of energy: from science and engineering to the analysis of energy policy. The series' scope includes but is not limited to, renewable and green energy, nuclear, fossil fuels and carbon capture, energy systems, energy storage and harvesting, batteries and fuel cells, power systems, energy efficiency, energy in buildings, energy policy, as well as energy-related topics in economics, management and transportation. Books published in LNE are original and timely and bridge between advanced textbooks and the forefront of research. Readers of LNE include postgraduate students and non-specialist researchers wishing to gain an accessible introduction to a field of research as well as professionals and researchers with a need for an up-to-date reference book on a well-defined topic. The series publishes single- and multi-authored volumes as well as advanced textbooks.

**Indexed in Scopus and EI Compendex** The Springer Energy board welcomes your book proposal. Please get in touch with the series via Anthony Doyle, Executive Editor, Springer (anthony.doyle@springer.com)

More information about this series at https://link.springer.com/bookseries/8874

Steven Beale · Werner Lehnert
Editors

# Electrochemical Cell Calculations with OpenFOAM

*Editors*
Steven Beale
Institute of Energy and Climate Research (IEK-14)
Forschungszentrum Jülich GmbH
Jülich, Germany

Mechanical and Materials Engineering
Queen's University
Kingston, ON, Canada

Werner Lehnert
Institute of Energy and Climate Research (IEK-14)
Forschungszentrum Jülich GmbH
Jülich, Germany

Faculty of Mechanical Engineering
RWTH Aachen University
Aachen, Germany

ISSN 2195-1284 ISSN 2195-1292 (electronic)
Lecture Notes in Energy
ISBN 978-3-030-92180-4 ISBN 978-3-030-92178-1 (eBook)
https://doi.org/10.1007/978-3-030-92178-1

© Springer Nature Switzerland AG 2022, corrected publication 2022
This work is subject to copyright. All rights are reserved by the Publisher, whether the whole or part of the material is concerned, specifically the rights of translation, reprinting, reuse of illustrations, recitation, broadcasting, reproduction on microfilms or in any other physical way, and transmission or information storage and retrieval, electronic adaptation, computer software, or by similar or dissimilar methodology now known or hereafter developed.
The use of general descriptive names, registered names, trademarks, service marks, etc. in this publication does not imply, even in the absence of a specific statement, that such names are exempt from the relevant protective laws and regulations and therefore free for general use.
The publisher, the authors and the editors are safe to assume that the advice and information in this book are believed to be true and accurate at the date of publication. Neither the publisher nor the authors or the editors give a warranty, expressed or implied, with respect to the material contained herein or for any errors or omissions that may have been made. The publisher remains neutral with regard to jurisdictional claims in published maps and institutional affiliations.

This Springer imprint is published by the registered company Springer Nature Switzerland AG
The registered company address is: Gewerbestrasse 11, 6330 Cham, Switzerland

# Preface

Computational fluid dynamics came of age as a stand-alone subject in the latter part of the 20th century following pioneering developments at Imperial College, and Los Alamos National Laboratory. Initial skepticism as to the applicability of general-purpose computer software to practical industrial applications, as well as novel and innovative scientific processes was quickly shown to be unfounded. Computational models can always provide the user qualitative insight into the physics governing natural and manufactured processes and devices. With skill and care, such models can also provide reliable quantitative information, *a priori*. Initially, engineers wrote CFD software in procedural languages such as FORTRAN or C, or adapted existing suites of software, typically based on what is referred to today, as the Finite Volume Method (FVM), and also the Finite Element Method (FEM). The argument for employing existing software/algorithms being that, other than for a very few specialized app developments, there was no real advantage, and no need to 'reinvent the wheel'. Tried and true methods, such as the SIMPLE algorithm developed by S. V. Patankar and D. B. Spalding, and many subsequent derivatives, are reliable and have already been highly optimized on a range of problems.

The analysis of electrochemical phenomena is an inter-disciplinary subject which may be considered a subset of the science of 'physicochemical hydrodynamics' a term coined by the Russian/Jewish scientist V. G. Levich to describe the interaction of physical and chemical phenomena with fluid mechanics (and vice versa). It is fair to say that the term CFD and not PCH has by-and-large caught on in the modern usage; nevertheless the conventional CFD literature is quite lacking with regard to this important subset of mathematical/engineering modeling. Fuel cells, electrolyzers, and batteries are all examples of PCH devices, where electrochemistry and transport phenomena proceed hand-in-hand. With renewed interest in global warming, decarbonization, renewable energy, and hydrogen as a clean energy currency, electrochemical processes have moved to the foreground for energy conversion. The need for reliable and open models was never greater.

A significant development at the turn of the twenty-first century was the creation, supply, and maintenance of open source modeling software. Of these OpenFOAM[1] enjoys widespread usage world wide, and especially in Germany, where the present writers are based. Like many other advances in CFD, OpenFOAM evolved out of the work of a group of researchers, such as H. Weller, H. Jasak, and many others, in the thermofluids group at Imperial College. The combination of open source software and the employment of object-oriented programming has proven to be highly successful and the code suite subsequently enjoyed exponential growth in development and application. In many industries it is a requirement that codes be open for the purposes of validation and verification. Moreover the open source paradigm is compatible with performance calculations on massively parallel computers from the perspectives of the licensing model, and numerical optimization. The application of open source software to modeling of electrochemical processes and devices is the subject of Annex 37 of the International Energy Agency Technology Collaboration Programme, Advanced Fuel Cells, which brings together leading researchers employing a variety of open source tools, from around the globe.

The goal of this project was to provide the reader with hands-on explanations of the basis for developing code specifically dedicated to electrochemical applications. All of the authors of this work have many years' experience, both in electrochemical engineering and also CFD/PCH with OpenFOAM and other software suites. The goal is to provide the reader with enough practical information that they can quickly start to develop their own models within the OpenFOAM environment. Since we cannot maintain suites of software, we invite the reader to write directly to the authors(s) should he or she require further information and/or access to specific source code. Indeed, it is hoped that eventually the above-mentioned IEA activities will lead to a suite of professional quality code being publically available/maintained by a community-of-users including both the authors and the readers of this book. The field is an exciting and creative one, which is rapidly advancing, as not only improved numerical techniques are created, but also the technology grows in application in society.

Jülich, Germany
October 2021

Steven Beale

---

[1] Reference to any product (commercial or otherwise) or corporation name is for information purposes only, and does not constitute recommendation or endorsement by the authors or editors.

The original version of the book was revised: The ESM material has been updated. The correction to the book is available at https://doi.org/10.1007/978-3-030-92178-1_10

# Contents

**Modeling of Electrochemical Cells** .... 1
Werner Lehnert and Uwe Reimer

**A Simple Electrochemical Cell Model** .... 21
Steven Beale

**Low-Temperature Polymer Electrolyte Fuel Cells** .... 59
Shidong Zhang

**High Temperature Polymer Electrolyte Fuel Cell Model** .... 87
Qing Cao

**Solid Oxide Fuel Cell Modeling with OpenFOAM®** .... 111
Dong Hyup Jeon

**Transport Modeling of High Temperature Fuel Cell Stacks** .... 125
Shidong Zhang and Robert Nishida

**Effective Transport Properties** .... 151
Pablo A. García-Salaberri

**Modeling Vanadium Redox Flow Batteries Using OpenFOAM** .... 169
Sangwon Kim, Dong Hyup Jeon, Sang Jun Yoon, and Dong Kyu Kim

**Liquid Metal Batteries** .... 193
Norbert Weber and Tom Weier

**Correction to: Electrochemical Cell Calculations with OpenFOAM** .... C1
Steven Beale and Werner Lehnert

# Modeling of Electrochemical Cells

**Werner Lehnert and Uwe Reimer**

## 1 Introduction

We are living in a very exciting time. The transition from fossil to regenerative energy sources is bringing about technological innovations, as well as societal changes. This leads us from a technological paradigm based on crude oil, natural gas, coal, and nuclear energy to one underpinned by regenerative power generation, such as that provided by wind turbines, solar cells, and hydroelectric power plants. The electricity generated from renewable sources can be fed directly into the power grid or, assuming the availability of appropriate technologies, stored in a suitable form. Hydrogen will play a key role as an energy carrier in this respect. Hydrogen can be produced from electricity and water by means of the electrolysis process, and can then be used directly or in subsequent processes, such as to produce synthetic fuels. The production of hydrogen, its storage and reconversion into electricity, for example through fuel cells, in both the stationary and transport sectors, has been receiving extensive attention in public and political discourse.

The storage of energy is vital due to the fluctuating nature of renewable power generation, and can be accomplished by various technologies. Batteries can be used to some degree. Examples are lead-acid, Li-ion, and liquid metal battery designs. An alternative is redox flow battery technology. If hydrogen is used as a storage medium, it can be stored in high-pressure vessels and large caverns, as well as in liquid form, and is then available not only for reconversion in fuel cells but also for use in other, more conventional conversion technologies. Hydrogen can also be

W. Lehnert (✉) · U. Reimer
Forschungszentrum Jülich GmbH, Institute of Energy and Climate Research, IEK-14, 52425 Jülich, Germany
e-mail: w.lehnert@fz-juelich.de

W. Lehnert
Modeling in Electrochemical Process Engineering, RWTH Aachen University, 52056 Aachen, Germany

© Springer Nature Switzerland AG 2022



S. Beale and W. Lehnert (eds.), *Electrochemical Cell Calculations with OpenFOAM*, Lecture Notes in Energy 42, https://doi.org/10.1007/978-3-030-92178-1_6

employed in mobile applications, with electrically-driven heavy-duty trucks powered by fuel cells and hydrogen seeming to be a future option. Battery-powered electrical vehicles, in addition to fuel cell-based hybrid vehicles for passenger cars, seem to be gaining wider acceptance.

In the 1990s, most automobile manufacturers devoted significant resources to the development of fuel cell technologies for passenger car applications. In most cases, other vehicle manufacturers were not regarded as competitors, but the main competition to fuel cell vehicles remained conventional vehicles with internal combustion engines. At this time, fuel cell vehicles were unable to compete with conventional systems due to factors of costs, service lifetimes, and insufficient hydrogen infrastructure. The same can be said for battery-powered passenger cars during this period. Trials showed that the technology worked, but neither the charging times needed for batteries and the driving ranges the vehicles allowed met customer requirements. In a sense, history repeated itself in this context. The first cars, back in the 1890s, were battery-powered; the internal combustion engine largely displaced this technology.

Today, the situation seems to be somewhat different. Battery technology has made many advances and there is increased environmental awareness. Local and global emissions must be reduced and the demand for emission-free vehicles to operate in densely-populated areas is becoming more strident. The field of mobile applications will probably feature a mixture of different technologies in the future. Electric vehicles for personal transport will likely draw their power from either batteries or hydrogen-fed fuel cells. The electric charging or hydrogen infrastructures will play an important role and must be made available for this technology to be successful. If this requirement is fulfilled, electrochemical converters and electrochemical storage systems will become more important in the future and gradually displace conventional combustion technologies. Electrochemistry, from electrocatalysis to applied electrochemistry and electrochemical process engineering, will gain importance in research, teaching, and in applications.

The following sections are not intended to be a reiteration of the fundamentals of electrochemistry or the complex physical/chemical relationships within electrochemical converters. Rather, it is intended to outline the role of the necessary parameters in the simulation of fuel cells, using the example of electrochemistry and transport in porous media.

## The Challenges of Electrochemistry

Electrochemical systems are omnipresent, even though they often perform their jobs inconspicuously. Batteries and accumulators appear in a wide variety of applications in everyday life. Every car, truck or motorcycle that utilizes an internal combustion engine contains a battery to start it that is intended to reliably operate for many years. Laptops, watches, smartphones, and many other everyday mobile devices receive their power from electrochemical converters. Every PC contains a small battery in the form of a button cell by which it can maintain a few necessary functions when the power supply is unavailable. Electrochemical converters and storage in the form of batteries, whether rechargeable or non-rechargeable, are an integral part of our everyday lives. However, battery storage systems are also utilized on a large scale to compensate for fluctuations in the power grid.

Significant, industrially-relevant and large-scale electrochemical systems include chlor-alkali electrolysis and fused-salt electrolysis. Electrochemical machining and electroplating are other typical methods used in industrial context. So-called sacrificial anodes on ship hulls, for example, also make use of electrochemical effects to protect metal hulls from corrosion.

As a result of the necessary sustainable energy transition, electrochemical energy converter and storage systems have become a major focus of research and development. In particular, fuel cells, electrolyzers, and batteries have become prominent. Thus, in addition to basic research, application-related research is also of great importance, with the main pillars of progress today being materials science, research on electrochemical mechanisms, and modeling and simulation.

The first evidence of electrochemical systems can be dated to the period from 250 BC to 250 AD. During excavations in Iraq, a copper cylinder and an iron rod were found in a concentric arrangement, insulated by a layer of asphalt. If one were to add an acid, this configuration would have resulted in a working battery. Although it has not been proven that the Parthians used this arrangement in the manner speculated upon, it may have been possible. The first proven electrochemical experiments date to the eighteenth century. Since that time, electrochemical research has continued to progress, even though interest in the subject has been subject to fluctuations. A few years before the turn of the millennium, despite industrial interest in fuel cells, interest in electrochemistry itself was fairly low. Wendt and Kreysa describe this in their 1999 book (Wendt and Kreysa 1999) on electrochemical engineering as follows:

> *"Electrochemistry soon became a very diversified field of – though still also today numerous – specialists working more and more on very particular problems, which are very often multidisciplinary in character, and it moved slowly to the rim of the physicochemist's vision. In the little justified view of the majority of the academic physicochemists who today imagine electrochemistry to be mainly manifested in terms of the Nernst equation and perhaps – if it goes that far – the Butler-Volmer equation, electrochemistry is now a chapter of science history."*
>
> (from: H. Wendt and G. Kreysa (1999)).

This pessimistic sentiment regarding electrochemistry in general has been overcome and electrochemical research is currently gaining momentum. However, a definitive definition of the term 'electrochemistry' probably does not exist. Electrochemistry is defined in different ways and depends strongly on the research area from which it is viewed. Adzic and Marinkovic write in the introduction to their book (Adzic and Marinkovic 2020):

> *"In general terms, electrochemistry is concerned principally with chemical reactions that are associated with the transfer of electrons or ions across interfaces. The most typical electrochemical interface is between a solid metal and a liquid solution of an electrolyte (ionic conductor), but any interface involving a predominantly electronic conductor and a predominantly ionic conductor is considered to be an electrochemical interface."*
>
> (from: R. Adzic and N. Marinkovic (2020)).

In his book (Holze 2009), Holze goes into greater detail regarding the interdisciplinary nature of electrochemistry:

> "*Electrochemistry is an extremely interdisciplinary area of science closely related to chemistry, physics, materials science, biology, surface science and a host of other fields. It is basically devoted to investigations of structures and dynamics as being present at interfaces between phases containing different types of mobile charged particles.*"
>
> (from: R. Holze (2009)).

The 'Electrochemical Dictionary' from 2008 (Bard et al. 2008) defines the topic as follows:

> "*Electrochemistry, as the name suggests, is a branch of chemical science that deals with the interrelation of electrical and chemical phenomena [i]. From the very beginning electrochemistry covers two main areas: the conversion of the energy of chemical reactions into electricity (electrochemical power sources) and the transformations of chemical compounds by the passage of an electric current (→ electrolysis). Electrochemistry is an interdisciplinary science that is mainly rooted in chemistry and physics; however, also linked to engineering and biochemistry/biology. According to this overlap, a simple definition cannot be given and it depends very much on the point-of-view what will be included. Electrochemistry is usually understood as part of physical chemistry, but it can be also subdivided according to the involved chemistry:*"
>
> (from: G. Inzelt, M. Lohrengel, F. Scholz in: A. J. Bard, G. Inzelt, F. Scholz 'Electrochemical Dictionary' (Bard et al. 2008)).

Probably the shortest, most general and comprehensive definition comes from Schmickler and Santos (2010), who do not describe electrochemistry in terms of specific disciplines or processes:

> "*Electrochemistry is the study of structures and processes at the interface between an electronic conductor (the electrode) and an ionic conductor (the electrolyte) or at the interface between two electrolytes.*"
>
> (from: W. Schmickler and E. Santos (2010)).

From these definitions, it is clear that electrochemistry cannot be considered a singular and clearly definable field. It is, rather, a multidisciplinary field.

### Scales in Electrochemistry

Electrochemistry is not only a multidisciplinary area but also has relevance at different scales. It is a multi-scale topic, starting with quantum mechanical effects and ending with applications in electrochemical engineering. Electrochemical reactions take place at surfaces and are linked to the transport processes of educts and their products. Electrocatalysis deals with atomic and molecular interactions, and is thus the smallest level in terms of the scale of its focus. The effects are described using quantum mechanical methods. At the next larger scale, phenomena in the nanometer to micrometer range are considered. For instance, if one considers electrodes in fuel cells, the properties of the porous stochastic structures, such as porosity, the distribution of pore radii, and also the three-phase zone as a function of the electrode morphology, all play a role. This scale can be considered a transition to applied electrochemistry. If, on the other hand, complete devices are considered, such as fuel cell or electrolysis stacks, as well as batteries, then process engineering in all its forms comes into play. This area can thus be considered *electrochemical process engineering*. The literature, as conveyed in books, discusses the topic in the field of fuel cells in particular. The entire spectrum of scales is covered in a number of publications [e.g., Larminie and Dicks 2003; Barbir 2005; Gasik 2008; O'Hayre et al. 2009; Kulikovsky 2010; Wang and Pasaogullari 2010; Franco 2013; Eikerling and Kulikovsky 2015; Hacker and Mitsushima 2018]. Interest in the topic of electrolysis is currently growing. The literature available on the subject is also steadily increasing, but has not (yet) reached the same volume as that on the topic of fuel cells [e.g., Bessarabov et al. 2015; Scott 2020; Godula-Jopek 2015].

## 1.1 Electrochemistry

The simulation of mass transport provides values for local mass distribution in the form of partial pressure, concentration, or mass or molar fractions within the framework of a model. On this basis, electrochemistry describes the rates of mass transfer as a function of current, which in turn are included in the transport equations as local sources and sinks for particular species. Already through the formulation of the electrochemical equations, fundamental assumptions are made that have a significant impact on the validity of the resulting model, and should be taken into account when interpreting the results. For example, the model of the polarization curve of a fuel cell is used, which is shown in Eq. (1):

$$E_{cell} = E_{Nernst} - \eta_{ohm} - \eta_{act} - \eta_{trans} \tag{1}$$

The cell voltage $E_{cell}$ is calculated on the basis of a thermodynamic reference state, the Nernst voltage $E_{Nernst}$, from which kinetic losses in the form of so-called overvoltages due to ohmic resistance, the activation overvoltage, and mass transport losses are subtracted. In the following, the contributions from the fields of thermodynamics and kinetics will be discussed.

**Thermodynamics**

Thermodynamics describes reactions in chemical equilibrium, as in Eq. (2) for the reaction of hydrogen and oxygen. This yields a first basic assumption, that the chemical equilibrium is established much faster than all transport processes:

$$H_2 + 0.5O_2 \leftrightarrow H_2O \tag{2}$$

This can be converted into an electric voltage with the aid of the definition of the electromotive force. The Nernst Eq. (3) expresses this voltage as a function of the local concentrations or partial pressure. As the logarithmic expression must not contain units, the molar fraction $X_i$ is used, which can be applied to both gas phases (as reduced partial pressure $X_i = p_i/p_i^0$) and aqueous solutions:

$$E_{Nernst} = \Delta E^0 + \frac{RT}{zF}\ln\left(\frac{X_{H2O}}{X_{H2}X_{O2}^{0.5}}\right) \tag{3}$$

At this point, another question arises: Should the water in the model be assumed to be produced in liquid or gaseous form? Physically, the answer is simple, as water will take the form of a liquid below the boiling point. However, in the context of the chosen transport model, it may well be more rational to always allow water to form in a gaseous state if, for example, the transport model does not represent a liquid phase. In fact, the following, different physical models can be used for the Nernst equation: "gas electrode," "three-phase boundary," and "flooded electrode" (Reimer et al. 2018). The latter two consider the presence of liquid water and are therefore only applicable to low-temperature fuel cell types, such as polymer electrolyte fuel cells (PEFCs). The resulting Nernst voltage differs by up to 200 mV. Compared to cell voltages from 0 to about 1 V, this difference seems quite large, at least initially. In purely practical terms, however, it has much less significance, as none of these models accurately represent the open cell voltage. Most models use an implicit correction that accurately determines the resulting cell voltage. Thus, the selection of the physical model of the Nernst voltage establishes the reference point for the overall model. According to Eq. (1), it should be noted that the absolute values of the kinetic parameters are not independent of this reference point when they are determined by, for instance, fitting an experimental polarization curve.

Another essential aspect of electrochemical cells is the spatial location of the reactions. They take place on the catalyst surfaces, which means that the corresponding half-cell reactions must be considered at each electrode (see Eq. (4)):

$$\begin{aligned} &Anode: H_2 \leftrightarrow 2H^+ + 2e^- \\ &Cathode: 2H^+ + 2e^- + 0.5O_2 \leftrightarrow H_2O \end{aligned} \tag{4}$$

The Nernst voltage of the overall reaction is obtained from the difference between the two half-cell potentials $E_{An}$ and $E_{Cat}$, as shown in Eqs. (5) and (6):

$$E_{An} = E_{An}^0 + \frac{RT}{2F}\ln\left(\frac{X_{H+}^2}{X_{H2}}\right) \tag{5}$$

$$E_{Cat} = E_{Cat}^0 + \frac{RT}{2F}\ln\left(\frac{X_{H2O}}{X_{H+}^2 X_{O2}^{0.5}}\right) \tag{6}$$

This shows that the local molar fraction of the protons has a direct influence on the value of the voltage. For models that represent the catalyst layer and the membrane as a two-dimensional surface, this influence is unimportant, as the identical value is simultaneously added and subtracted by the logarithm. However, if the membrane and catalyst layer are spatially-resolved, the resulting gradients of the proton concentration cannot be directly neglected.

From Eqs. (3), (5), and (6), it follows as a further assumption that only the species from the reaction equation determine the resulting potential difference. The material of the catalyst has no influence on it. This is correct for systems in which the catalyst layers for the anode and cathode are the same and do not change during operation. In technical systems, however, different combinations for the anode and cathode are often used. Although pure Pt is often used on the hydrogen side, PtRu alloys are also found in PEFC systems. In PEM electrolysis, the catalyst on the oxygen side is usually $IrO_2$, or sometimes a mixture of $IrO_2/RuO_2$. It is often very difficult to assess whether, and to what extent, these material combinations contribute to the potential difference. This fact is mostly neglected in the context of modeling, as the selection of the model parameters are typically based on experimentally-determined polarization curves (see the discussion on the Nernst voltage). It should be clarified in advance whether this assumption is valid for the operating range under consideration. A good example is a PEFC with Pt catalysts on the anode and cathode, wherein identical electrodes can be assumed, neglecting the different structure and loading. This is valid over almost the entire operating range. Only in the vicinity of the open cell voltage does partial oxidation of the Pt occur on the cathode, which can result in a potential shift of up to 50 mV (Reimer et al. 2019; Cai 2019). In addition to the oxidation of the surface, the structure of the catalyst particles, which are only a few nanometers in size, also has an influence. For a more detailed discussion, please refer to the work by Holdcroft (2014) and book by Eikerling and Kulikovsky (2015), which contain a comprehensive overview of the effects described.

**Kinetics**

As a result of the thermodynamic considerations, the voltage is obtained as the potential difference. The voltage obtained in this manner should correspond to the open cell voltage. However, the presence of a significant electrical current requires the occurrence of directed transport processes, which necessitate a certain amount of the available energy. Electrochemical kinetics describe this loss as a so-called "overvoltage." In a highly simplified way, two essential transport processes can be identified that occur for all electrochemical reactions on charged surfaces. On the one hand, there is a transition of electrons from the conduction band of the catalyst through the metal surface into the valence band of adsorbed species. On the other, it entails the transport of charged species through the electrochemical double layer on the catalyst surface. These two effects are described by the Butler-Volmer Eq. (7). Equation (7) contains two exponential terms, the first describing the forward reaction and the second the backward one. This is especially important near electrochemical equilibria. For technically relevant current densities, however, the reverse reaction is of no significance and Eq. (7) simplifies, as an exact mathematical limit, consideration to the Tafel Eq. (8) (Reimer et al. 2018; Hamann and Vielstich 2005; Bard and Faulkner 1988):

$$j = j_0 \left( \frac{c_i}{c_i^{ref}} \right)^{\gamma} \left[ \exp\left( \frac{\alpha n F}{RT} \eta_{act} \right) - \exp\left( \frac{(1-\alpha) n F}{RT} \eta_{act} \right) \right] \tag{7}$$

$$j = j_0 \left( \frac{c_i}{c_i^{ref}} \right)^{\gamma} \left[ \exp\left( \frac{\alpha n F}{RT} \eta_{act} \right) \right] \tag{8}$$

There are far more complex models for - electron transfer in electrochemical reactions, such as Marcus theory (Hamann and Vielstich 2005; Bard and Faulkner 1988), for which the 1992 Nobel Prize in Chemistry was awarded. Nevertheless, Eqs. (7) and (8) have become established as a standard. A major reason for this may be the accessibility of the required parameters. Equations (7) and (8) include six model parameters, in addition to the temperature T, the universal gas constant R, and the Faraday constant F. Based on theoretical considerations, the exchange current density $j_0$ corresponds to a rate constant for the reaction at the surface. Therefore, the parameter $j_0$ can be related to the size of the active catalyst surface. The reaction rate also depends on the local concentration of the reacting species at the surface. This is described by the concentration $c_i$. The reference concentration $c_i^{ref}$ is the concentration at which the value for $j_0$ was originally measured. In turn, the parameter $\gamma$ is called the reaction order and is practically used as a fitting parameter. However, assumption $\gamma = 1$ is commonly used, as the model approach is already highly simplified and describes, to be precise, only a 1-electron transition. In turn, the parameter $\alpha$ is called the symmetry factor or transfer coefficient. In a broader sense, properties of the electrochemical double layer can be assigned to it. A change in the thickness of the double layer or, for instance, the material of the catalyst surface, should therefore have an effect on this parameter. In fact, a different value of the Tafel slope can be experimentally-determined in the region in which the Pt surface is partially oxidized (Song and Zhang 2008; Wakabayashi 2005; Stassi 2006). The parameter n is included for formal reasons and describes the number of electrons exchanged in the rate-determining step, and should not be confused with the parameter z from Eq. (3). For the Tafel equation, in its linear representation as $\eta = f(\ln j)$, the product $\alpha \cdot n$ can be determined from the slope and the value for $j_0$ from the intercept. Thus, the activation overvoltage may be calculated from two terms ($\alpha \cdot n$ and $j_0$) if the local concentration is given.

Models that represent the electrodes as a two-dimensional interface often utilize the physical model of a "combined electrode" (see Fig. 1). Here, the anode and cathode are combined to form an effective electrode. This is always an elegant solution if the system has a hydrogen electrode on one side, which can thus be treated as a reference electrode.

The great advantage of this method is that the activation overvoltage can be calculated with only two effective parameters, which are directly accessible from the characteristic curve. The limits of this method are reached when other reactions occur at the hydrogen electrode or deactivation of the catalyst takes place, such as when operating a PEFC with reformate. These additional effects must then be taken into account as a "disturbance" through parameters. Another possibility is to consider two half-cells and use two Tafel equations analogous to Eqs. (5) and (6). As an additional challenge, however, the kinetic parameters must be determined separately for each electrode. One possibility is to use reference electrodes in the membrane.

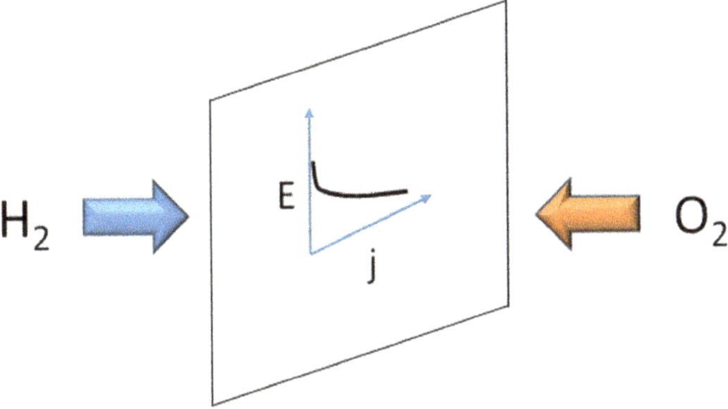


**Fig. 1** Model of the "combined electrode" with a two-dimensional approximation of the two catalyst layers

Unfortunately, an absolute determination of each contribution is hardly possible, because the proton concentration cannot be controlled and the reference potential shifts according to Eq. (5) depending on the current density, with the positioning also having a direct influence on the measurement (Liu et al. 2004; Kulikovsky and Berg 2015). Despite these difficulties, this approach offers a direct experimental approach for validating the model parameters.

Another possibility is to represent the catalyst layer in 3D, as is shown in Fig. 2. However, this requires knowledge of the internal structure. As the simplest approach, a homogeneous structure can be assumed, with different gradients arising in the layer depending on the parameters selected.

It can be seen from the illustration in Fig. 2 that for this purpose, the catalyst layers on both sides must be considered separately. However, it can also be assumed that the contribution of the anode can be neglected when operating with pure $H_2$. The advantage of this type of model is that it links the electrochemical conversion

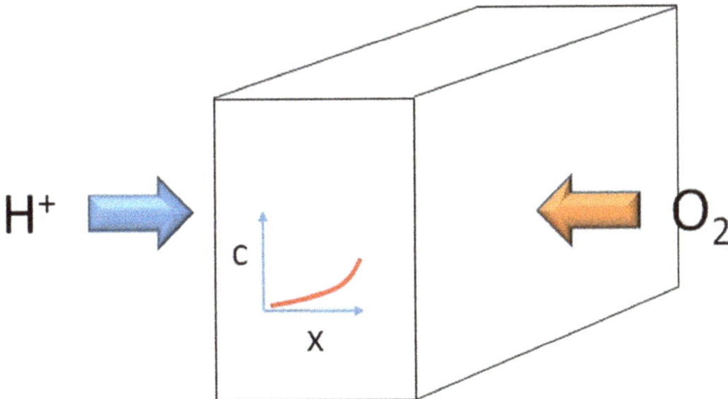


**Fig. 2** Model of the "homogeneous catalyst layer" (cathode)

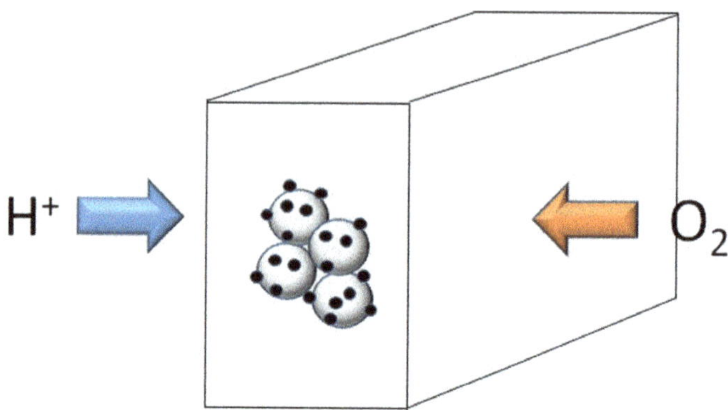


**Fig. 3** "Agglomerate model" with illustrated structure of the catalyst layer

with transport equations for protons and gas species. Thus, statements regarding the degree of utilization of the electrodes as a function of their loading and layer thickness are possible (Kulikovsky 2010a, b; Perry et al. 1998; Boyer et al. 2000). However, the assumption of a homogeneous catalyst layer contradicts the internal structure of the catalyst layer, which is known from the manufacturing process and analysis of the electrode morphology. This structure can be taken into account by drawing on the so-called agglomerate models (Sun et al. 2005; Sousa et al. 2010; Kamarajugadda and Mazumder 2012; Xing and Mamlouk 2013) (see Fig. 3).

This type of model takes into account the spatial distribution of the platinum particles and their connection to the carbon carrier material and polymer phase. The increased accuracy in the geometric description, however, in turn requires a larger number of model parameters that must be determined via suitable experiments. As an example, the eleven most important parameters required in the model of Sun et al. (2005) are listed below:

- Exchange current density.
- Symmetry factor.
- Pt loading.
- Pt particle diameter.
- Effective Pt surface ratio.
- Radius of the agglomerate.
- Effective specific agglomerate surface area.
- Catalyst layer thickness.
- Porosity of the catalyst layers.
- Thickness of the electrolyte film covering each agglomerate.
- Electrolyte fraction in the agglomerate.

In general, it can be said that the simulation of a "correct characteristic" at different operating conditions is at least a first indication of the correct model description. However, reproducing the experimentally-obtained polarization curve by means of the model is a necessary but not a sufficient condition for the model's validation. For instance, experiments on high-temperature PEFCs in co-flow and counter-flow result in identical polarizations curves within the measurement accuracy, but the temperature distribution and even current density distribution clearly differ (Lüke et al. 2012). The following three points describe the fundamental assumptions that have a key influence on the simulation results and should therefore be taken into account when interpreting the results:

- The establishment of the thermodynamic equilibrium is much faster than all transport phenomena.
- The choice of the electrode model pertaining to the Nernst voltage affects the absolute values of the kinetic parameters.
- The electrode material does not change during operation and is the same on both sides of the electrochemical cell.

When evaluating experiments in order to determine parameters, the same assumptions should be made as in the modeling. Only then can the parameters be meaningful within the framework of the models used.

## 1.2 Mass Transport in Porous Media

Low-temperature, as well as high-temperature fuel cells and electrolyzers, consist to a large extent of stochastic porous media. The typical porous components of a PEFC include the gas diffusion layers (GDLs), the microporous layers, the electrodes, and, in a broader sense, the membrane. Oftentimes, the only "deterministic" components are the bipolar plates with the incorporated flow channels, with both the ribs and channels of the bipolar plates having an interface with the porous GDLs on one side. From this simple consideration, it is clear that porous media exert a significant influence on fuel cell behavior and not only in terms of mass, heat, and charge transport but also in relation to electrochemical conversion in the electrodes.

This will be outlined using the example of mass transport in the gas diffusion layer. The transport of a fluid in a porous medium is disturbed by the solid components. These limit the free volume through which the fluid can be transported. The porosity, i.e., the ratio of the void volume to the total volume, decreases with increasing solid content. In addition to the effect of porosity, other microstructural properties have an influence on effective transport. The microstructure can lead to the blocking of the straight pathways for the fluid through the medium, increasing the length of its pathway. This is accounted for by the tortuosity. In addition, constrictions on the transport paths can occur with the result that the pore radii along the transport paths change or vary from pore to pore. Furthermore, the pores or transport paths may be interconnected. In addition, dead end pores can be created by the solid fraction, as is shown in Fig. 4a, top right.

In these three-dimensional, stochastic porous media, transport must be described using mathematical models. This can be performed in several ways, and it is often convenient to simplify the model's spatial geometry. For instance, the pores can be

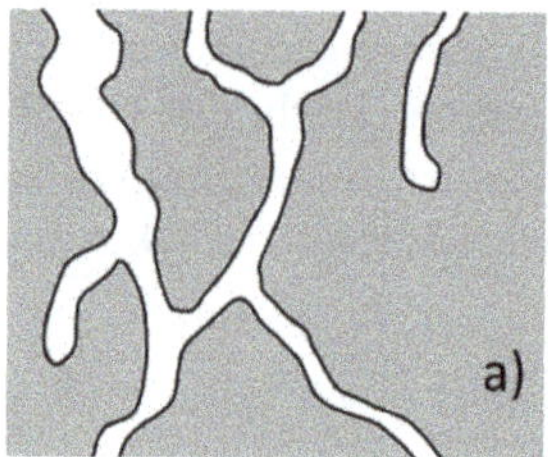

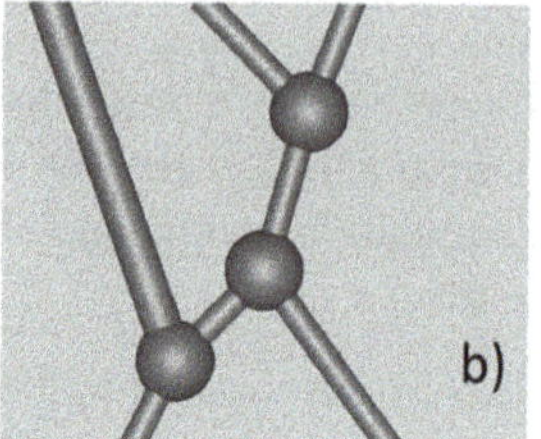

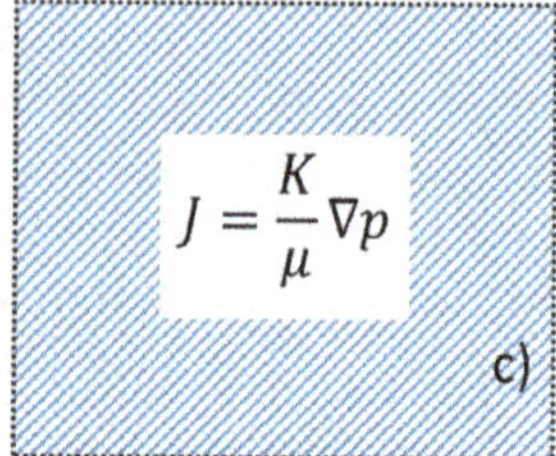


**Fig. 4** Sketch of the cross-section through a porous medium: **a** real structure with dead-end pores; **b** abstracted model of the pore network; and **c** representation as representative elementary volume with average properties (according to Eq. 14)

described by a size distribution and the interconnection of all pores represented by a network (Sahimi 2011), as is depicted in Fig. 4b.

It would, in principle, be straightforward to solve the transport equations in the spatially-resolved structures or stochastic model structures that are equivalent to the real structures in a statistical sense. This would require a detailed knowledge of these underlying structures. With current tomography and image-processing methods, these structures can be detected and presented in a manner suitable for transport simulations (Sahimi 2011; Thiedmann et al. 2008, 2009; Wang et al. 2010). An alternative to describe the transport properties in porous media are pore network models, which are an abstraction of the real structures and feature equivalent properties to the real materials in the statistical sense. In these networks, fluid transport is often described using simplified transport models (Gostick et al. 2007; Markicevic et al. 2007; Agaesse 2016). In general, this can greatly reduce the computational resources required while maintaining good accuracy among the transport parameters, as was recently shown by Gackiewicz et al. (2021) in a direct comparison of two methods for flow in sand layers.

In contrast, in terms of cell and stack simulation, continuum models are used to describe transport phenomena in porous structures. The two aforementioned approaches, as successfully as they can be used, require unacceptably high computation times and core memory when large volumes/regions are to be simulated, such as in the case of complete cells or stacks. When statistical properties are averaged out based on the considered representative elementary volume and local effects thereby play a minor role, volume-averaged models are a suitable choice. Instead of the knowledge of detailed structures of the porous media, consistently effective parameters are required to characterize the single-phase and multi-phase transport processes. As an example, consider the determination of the effective diffusion coefficient $D_{eff}$ for flow-through of a porous medium. The value of $D_{eff}$ depends on the properties of the fluid, as well as on the porous medium's structure. Considering bulk diffusion, the parameter multiplied by the diffusion coefficient includes, in the simplest case, the porosity $\varepsilon$ and the tortuosity $\tau$. For the effective diffusion coefficient, $D_{eff}$ is obtained as:

$$D_{eff} = \frac{\varepsilon}{\tau} D_{bulk} \tag{9}$$

A reduction in the porosity ε or an increase in the transport path (always $\tau > 1$) leads to a reduction in the effective diffusion coefficient.

Porosity data are available for many materials, but care should be taken as to which definition of porosity is meant. Thus, for mass transport through a porous layer (e.g., through a GDL), only the fraction that enables gas transport through the layer from one side to the other is relevant. However, dead-end pores with a connection to the surface but without continuous paths (see Fig. 4a) have no relevance in terms of mass transport. Dead-end pores inside the porous medium connected to transport-relevant pores also do not contribute to mass transport via the porous layer. In the field of catalysis, or electrocatalysis, these pores are of importance. Closed pores are also irrelevant in terms of mass transfer. They have an influence on the mechanical, thermal behavior and, if applicable, the electrical conductivity. Porosity data from the literature should therefore be critically considered. If the method of porosity measurement is known, the measured porosity can be assigned. Porosity and other characteristic data for different materials, such as GDLs, can be found in the relevant literature (Zamel and Li 2013; Ozden et al. 2019; Yang et al. 2021; Yuan et al. 2021; https: 2021).

Another aspect is the tortuosity, which is often defined in an intuitive manner as being the quotient of the effective path length $L_{eff}$ and thickness of the material $L_{short}$:

$$\tau = \frac{L_{eff}}{L_{short}} \tag{10}$$

This definition refers to the geometric path length, i.e., the shortest paths between opposing surfaces. However, it does not take into account the influences of wide and narrow pores on mass transfer. In this regard, a distinction must be made between geometric and transport-based tortuosity, both values of which can significantly differ. In the literature, tortuosity is often described as a material constant which, in combination with porosity, diminishes the effective mass transport in the porous medium. However, it has been shown that tortuosity is not a pure structural parameter and is a function of the temperature, transported species, pressure, and possible reactions within the porous structure and flow conditions (Ghanbarian et al. 2013; Keil 1999). This flow-based tortuosity is a variable that relates to random porous media and cannot be measured directly.

An alternative definition that results in more realistic values is presented in Eq. (11) (a discussion of the differences between the two definitions can be found in a study by Epstein (1989), where the terms tortuosity and tortuosity factor are discussed):

$$\tau = \left[\frac{L_{eff}}{L_{short}}\right]^2 \tag{11}$$

In order to take into account the variable pore diameter along the diffusion path, the constrictivity factor δ is sometimes used in the literature, which is intended to represent the hindrance of the fluid and lead to smaller values for $D_{eff}$, as indicated in Eq. (12) (Coutelieris and Delgado 2012):

$$D_{eff} = \frac{\delta \cdot \varepsilon}{\tau} D_{bulk} \tag{12}$$

Regardless of the descriptions or underlying models, it is necessary to use suitable parameters to describe the mass transport. In principle, it is irrelevant whether these parameters are determined experimentally (Tjaden et al. 2018) or via detailed simulations in real structures (Wang et al. 2010; Froning et al. 2013). Due to the uncertainty of parameter determination in real porous structures, tortuosity is often taken as a fitting parameter. Alternatively, different empirical correlations can be found in the literature, but refer to specific underlying structures. For instance, the Bruggeman correction is mentioned here (Eq. 13) (Markicevic et al. 2007; Bruggeman 1935). This model was originally developed for media consisting of regular stacks of spheres. Its validity to fuel cell materials, however, is questionable (Yuan et al. 2021). Further correlations can be found in the relevant literature (Sahimi 2011; Tjaden 2018).

$$D_{eff} = \varepsilon^{1.5} D_{bulk} \tag{13}$$

Diffusion in the porous GDLs is typically superimposed by means of convective mass transport. In this respect, permeability through the porous media is of additional importance and must be considered in the models. Due to the small Reynolds number, a Darcy approach can typically be applied in the GDLs (Eq. 14):

$$J = \frac{K}{\mu} \nabla p \tag{14}$$

where J is the superficial velocity that is caused by the pressure gradient $\nabla p$. Furthermore, K is the permeability and μ the dynamic viscosity. Permeability also depends on the microstructure of the materials and is measured through-plane in many cases, when this is possible using standard methods. Comprehensive through-plane and in-plane data for typical materials are only partially available. It is common to use these measured permeabilities in the models, even when it is possible to calculate permeability via effective structural parameters within the framework of models, as is done, for example, in the mean transport pore model (Arnost and Schneider 1995). However, a lack of measured or calculated parameters for the application is also apparent here. In this respect, it is understandable and acceptable to employ effective measured or calculated permeability.

Thus, it becomes apparent that several parameters are necessary for the description of the single-phase mass transport in the framework of the continuum models which, however, are only partially available and feature large uncertainties. In a typical PEFC, liquid water is also present. Therefore, the phenomena of two-phase transport must be implemented in the models. Models of this are presented in the subsequent chapters of this book. In this section, only the available and, in particular, the unavailable experimental and validated parameters will be briefly highlighted. Many of the models were originally developed for the fields of groundwater flow and petroleum engineering (Bear 1988), and have been applied therein with great success. An essential difference from typical fuel cell materials is that in the field of petroleum engineering, there are large areas with comparable dimensions in all spatial directions when performing experiments for the determination of parameters. In contrast, porous components with large x–y but small z expansions are found in fuel cells. In this respect, near-surface effects in the z-direction dominate in thin fuel cell materials, in contrast to the materials utilized in petroleum engineering. In the absence of specific models and parameters adapted to typical fuel cell models, established models and data from the petroleum engineering domain are often used.

In order to describe two-phase transport in the porous fuel components in the context of continuum models, the capillary pressure saturation curve and relative permeabilities are essential considerations. In most PEFC models, the Leverett-Udell function is used as the capillary pressure–saturation relationship (Leverett 1941; Udell 1985). Gostick et al. showed that the measured capillary pressure saturation curves of typical GDL materials with varying hydrophobicities differ in shape (Gostick et al. 2012; Gostick et al. 2010). It should also be noted that Udell's fit does not adequately reflect the underlying experimental data (Si et al. 2015). Furthermore, the data do not relate to typical fuel cell materials. Despite these two factors, the Leverett-Udell function is widely used. Another aspect made evident in the experimental data is the hysteresis of the capillary pressure saturation curve. To the best of the authors' knowledge, this has not been discussed in the context of continuum models describing fuel cell behavior.

If one considers the data with respect to relative permeability or to structural parameters relevant for material transport as a function of water saturation, a comparable picture emerges; little experimental data is currently available for typical fuel cell materials.

In actual fuel cells, the MEA components are compressed to ensure a low contact resistance between the ribs of the flow-field and GDL. On the other hand, the compression must not be so strong as to hinder mass transfer below the ribs. Typically, the compression is about 30%, with the GDL being the component subject to the highest compression. Thus, compressed and non-compressed regions and transition regions may be found in a cell. It is apparent that compression combined with the anisotropy of the GDL materials does not make the situation easier. Experimental data describing these effects in relation to the necessary structural parameters or directly to the mass transport are also sparse regarding the typical fuel cell and electrolyzer materials. For this reason, transport simulations in tomographically-examined materials are increasingly employed (Hoppe 2021). Typical data and parameters that consider the anisotropy of the materials and compression are:

- Porosity
- Tortuosity

- The capillary pressure saturation curve
- Relative permeabilities

It does not matter whether the data originate from experiments or validated simulations. Rather, the decisive factor is that the data are consistent across the respective material classes. Naturally, increasing the complexity of the models would require more and better data. At this point, the data should not be weighted in the sense of a sensitivity analysis, although not all parameters have the same impact on the simulation result; some parameters can also be viewed as fitted values.

## 1.3 Conclusions

Although they may seem simple at first sight, fuel cells present highly complex scientific and technical challenges. This is also true in the field of modeling and simulation within the framework of continuum models, which was briefly touched upon here. It turns out that models for the description of fuel cell behavior are available, and these generally work well in the context of their intended applications. Unfortunately, it must be stated that it is not easy to obtain the necessary parameters for the mentioned models. More intensive collaboration between experimentalists and modelers thus seems to be necessary, whereby the synergies may become more effective and the mutual benefits more obvious, as was demonstrated, e.g., in Beale et al. (2021). However, experience shows that it is not always easy to engage in such cooperative endeavors.

In the field of modeling and simulation, multi-physics and multiscale models have been developed to describe fuel cell behavior. In recent years, CFD simulations have proven to be of particular importance in this regard. CFD simulations enable the modeling of complex interactions within a cell or a stack. For a long period of time, commercial CFD tools have dominated the simulation community. This has certainly contributed to the popularity of fuel cell simulation, and it has seemed easy to perform meaningful model calculations with these tools, provided that the appropriate computer infrastructure has been in place. However, it is apparent that many of the published results are only partially validated and difficult to interpret in their complexity. In this regard, it seems useful to make a few general statements with respect to modeling and simulation, as per the books by Ingham et al. (2000) and Moeller (2004).

Even if it is apparently trivial, it should be noted that before modeling work is commenced, its goals should be determined. The level of detail of a model, as well as its balance region, result from the target's definition. Aside from the demand for a clear definition of the goal, the systematic approach to modeling and simulation is essential, as the above-mentioned authors make clear. It is recommended not to start with the most complex model, but to successively increase the complexity and check all intermediate steps. It should also be ensured that the necessary model parameters, for example material data, are available. These data should derive from appropriate experiments or simulations. This requirement, which is in fact often taken for granted, is more complex and difficult to meet than the previous ones. In addition, it is self-evident that a model must be validated in order to be usefully applicable. The first step to this is to ensure that it corresponds to basic physical and chemical understanding. An experimental validation via—appropriate—experiments is to be aimed for in any case. If this is not possible, the model should be compared with independently-developed ones. Unfortunately, these validation steps are complex and sometimes highly time-consuming. Furthermore, both parameter determination and validation require close collaboration among experimental working groups. Unfortunately, the authors' experience shows that such cooperation sometimes proves to difficult due to different understandings and requirements of the experiments or simulations. In this regard, there is demand to build up a mutual understanding and value the work in an appropriate manner. Only then can the synergy of experiments and simulations be successfully executed.

The following chapters include models that are accessible to everyone. This accessibility is an advantage of open source software over commercial tools. The models presented in the subsequent chapters describe modeling approaches in detail and are comprehensible due to these descriptions. In addition, further necessary literature is outlined. Thus, models are available that can be applied to a wide range of applications and further developed to adapt them to the respective goals of modeling and simulation.

## References


https://fuelcellsetc.com/2013/03/comparing-gas-diffusion-layers-gdl-/ (Access May 2021)

Adzic R, Marinkovic N (2020) Platinum monolayer electrocatalysts, Springer Nature Switzerland AG

Agaesse MT (2016) Simulations of one and two-phase flows in porous microstructures, fromtomographic images of gas diffusion layers of proton exchange membrane fuel cells. PhD Thesis, University of Toulouse, France

Arnost D, Schneider P (1995) Dynamic transport of multicomponent mixtures of gases in porous media. Chem Eng J 57(91)

Barbir F (2005) PEM fuel cells. Elsevier, Amsterdam

Bard AJ, Faulkner LR (1988) Electrochemical methods - fundamentals and applications, 2nd edn. Wiley, New York

Bard AJ, Inzelt G, Scholz F (2008) Electrochemical dictionary. Springer

Beale SB, Andersson M, Boigues-Muñoz C, Frandsen HL, Lin Z, McPhail SJ, Ni M, Sundén B, Weber A, Weber AZ (2021) Continuum scale modelling and complementary experimentation of solid oxide cells. Prog Energy Combust Sci 85:100902

Bear J (1988) Dynamics of fluids in porous media. Dover Publication

Bessarabov D, Wang H, Li H, Zhao N (2015) Electrolysis for hydrogen production - principles and applications. CRC Press

Boyer CC, Anthony RG, Appleby AJ (2000) J Appl Electrochem 30:777–786

Bruggeman DAG (1935) Calculation of different physical constants of heterogeneous substances, dielectric constant and conductivity of media of isotropic substances. Ann Phys Ser-5 24:636–664

Cai Y (2019) PhD thesis. RWTH Aachen University, Germany

Coutelieris FA, Delgado JMPQ (2012) Transport processes in porous media. Springer, New York
Eikerling M, Kulikovsky AA (2015) Polymer electrolyte fuel cells — physical principles of materials and operation. CRC Press, Boca Raton
Epstein N (1989) On tortuosity and the tortuosity factor in flow and diffusion through porous media. Chem Eng Sci 44:777–779
Franco AA (ed) (2013) Polymer electrolyte fuel cells - science, applications, and challenges. CRC Press, Boca Raton
Froning D, Brinkmann J, Reimer U, Schmidt V, Lehnert W, Stolten D (2013) 3D analysis, modeling and simulation of transport processes incompressed fibrous microstructures, using the Lattice Boltzmann method. Electrochim Acta 110:325–334
Gackiewicz B, Lamorski K, Sławiński C, Hsu S-Y, Chang L-C (2021) An intercomparison of the porenetwork to the Navier-Stokes modeling approach applied for saturated conductivity estimation from X-ray CT images. Sci Rep 11:5859
Gasik M (2008) Materials for fuel cells. CRC Press, Boca Raton
Ghanbarian B, Hunt AG, Ewing RP, Sahimi M (2013) Tortuosity in porous media: a critical review. Soil Sci Soc Am J 77:1461
Godula-Jopek A (2015) Hydrogen production. Wiley-VCH
Gostick JT, Ioannidis MA, Fowler MW, Pritzker MD (2007) Pore network modeling of fibrous gas diffusion layers for polymer electrolyte membrane fuel cells. J Power Sources 173:277–290
Gostick JT, Fowler MW, Pritzker MD, Loannidis MA (2012) Porosimetry and characterization of the capillary properties of gas diffusion media. In: Wang H, Yuan X-Z, Li H (eds) PEM fuel cell, diagnostic tools. CRC Press, Taylor & Francis Group, LLC
Gostick JT, Loannidis MA, Fowler MW, Pritzker MD (2010) Characterization of the capillary properties of gas diffusion media. In: Pasaogullari U, Wang C-Y (eds) Modeling and diagnostics of polymer electrolyte fuel cells, modern aspects of electrochemistry, vol 49. Springer Science+Business Media, LLC
Hacker V, Mitsushima S (2018) Fuel cells and hydrogen — from fundamentals to applied research. Elsevier
Hamann CH, Vielstich W (2005) Elektrochemie, 2nd edn. Wiley-VCH, Weinheim
Holdcroft S (2014) Chem Mater 26:381–393
Holze R (2009) Surface and interface analysis — an electrochemists toolbox. Springer
Hoppe E (2021) Dissertation RWTH Aachen University
Ingham J, Dunn IJ, Heinzle E, Přenosil JE, Snape JB (2000) Chemical engineering dynamics. Wiley-VCH
Kamarajugadda S, Mazumder S (2012) J Power Sources 208:328–339
Keil F (1999) Diffusion und Chemische Reaktion in der Gas/Feststoff-Katalyse. Springer
Kulikovsky AA (2010a) Analytical modelling of fuel cells. Elsevier, Amsterdam
Kulikovsky AA (2010b) Electrochim Acta 55:6391–6401
Kulikovsky AA, Berg P (2015) J Electrochem Soc 162(8):F843–F848
Larminie J, Dicks A (2003) Fuel cell systems explained, 2nd edn. Wiley, Chichester
Leverett MC (1941) Capillary behavior in porous solids. Soc Pet Eng 142:152–169
Liu Z, Wainright JS, Huang W, Savinell RF (2004) Electrochim Acta 49:923–935
Lüke L, Janßen H, Kvesić M, Lehnert W, Stolten D (2012) Int J Hydrogen Energy 37:9171–9181
Markicevic B, Bazylak A, Djilali N (2007) Determination of transport parameters for multiphase flow in porous gas diffusion electrodes using a capillary network model. J Power Sources 171:706–717
Moeller DPF (2004) Mathematical and computational modeling and simulation. Springer
Ozden A, Shahgaldi S, L* X, Hamdullahpur F (2019) A review of gas diffusion layers for proton exchange membrane fuel cells-With a focus on characteristics, characterization techniques, materials and designs. Prog Energy Combust Sci 74:50–102
O'Hayre RP, Cha S-W, Colella WG, Prinz FB (2009) Fuel cell fundamentals, 2nd edn. Wiley, New York
Perry ML, Newman J, Cairns EJ (1998) J Electrochem Soc 145:5–15

Reimer U, Cai Y, Li R, Froning D, Lehnert W (2019) J Electrochem Soc 166(7):F3098–F3104
Reimer U, Lehnert W, Holade Y, Kokoh B (2018) Irreversible losses in fuel cells. In: Hacker V, Mitsushima S (eds) Fuel cells and hydrogen - from fundamentals to applied research. Elsevier
Sahimi M (2011) Flow and transport in porous media and fractured rock — from classical methods to modern approaches, 2nd edn. Wiley-VCH, Berlin
Schmickler W, Santos E (2010) Interfacial electrochemistry, 2nd edn. Springer, Berlin
Scott K (2020) Electrochemical methods for hydrogen production. The Royal Society of Chemistry
Si C, Wang X-D, Yan W-M, Wang T-H (2015) A comprehensive review on measurement and correlation development of capillary pressure for two-phase modeling of proton exchange membrane fuel cells. J Chem 876821
Song C, Zhang J (2008) Electrocatalytic oxygen reduction reaction. In: Zhang J (ed) PEM fuel cell electrocatalysts and catalyst layers. Springer
Sousa T, Mamlouk M, Scott K (2010) Chem Eng Sci 65:2513–2530
Stassi A (2006) J Appl Electrochem 36:1143
Sun W, Peppley BA, Karan K (2005) Electrochim Acta 50:3359–3374
Thiedmann R, Fleischer F, Hartnig C, Lehnert W, Schmidt V (2008) Stochastic 3D modeling of the GDL structure in PEMFCs based on thin section detection. J Electrochem Soc 155(4):B391–B399
Thiedmann R, Hartnig C, Manke I, Schmidt V, Lehnert W (2009) Local structural characteristics of pore space in GDLs of PEM fuel cells based on geometric 3D graphs. J Electrochem Soc 156(11):B1339–B1347
Tjaden B, Brett DJL, Shearing PR (2018) Tortuosity in electrochemical devices: a review of calculation approaches. Int Mater Rev 63(2):47–67
Udell KS (1985) Heat transfer in porous media considering phase change and capillarity-the heat pipe effect. Int J Heat Mass Transf 28(2):485–495
Wakabayashi N (2005) J Electroanal Chem 574:339
Wang Y, Cho S, Thiedmann R, Schmidt V, Lehnert W, Feng X (2010) Stochastic modeling and direct simulation of the diffusion media for polymer electrolyte fuel cells. Int J Heat Mass Transfer 53:1128–1138
Wang C-Y, Pasaogullari U (eds) (2010) Modeling and diagnostics of polymer electrolyte fuel cells. Springer, New York
Wendt H, Kreysa G (1999) Electrochemical engineering — science and technology in chemical and other industries. Springer
Xing L, Mamlouk M, Scott K (2013) Energy 61:196–210
Yang Y, Zhou X, Li B, Zhang C (2021) Recent progress of the gas diffusion layer in proton exchange membrane fuel cells: Material and structure designs of microporous layer. Int J Hydrogen Energy 46:4259–4282
Yuan X-Z, Nayoze-Coynel Ch, Shaigan N, Fisher D, Zhao N, Zamel N, Gazdzicki P, Ulsh M, Friedrich K, Girard F, Groos U (2021) A review of functions, attributes, properties and measurements for the quality control of proton exchange membrane fuel cell components. J Power Sources 491:229540
Zamel N, Li X (2013) Effective transport properties for polymer electrolyte membrane fuel cells – With a focus on the gas diffusion layer. Prog Energy Combust Sci 39:111–146

# A Simple Electrochemical Cell Model

**Steven Beale**

**ABSTRACT** A case corresponding to the simplest possible electrochemical cell, namely a single oxygen electrode for a polymer electrolyte electrochemical cell, with constant properties, is presented herein. The target audience is one that is familiar with the computer code Open Source Field Operation and Manipulation (OpenFOAM), with little or no knowledge of electrochemistry, who wish to enter the field, as well as those with some knowledge of electrochemistry who wish to build their first mathematical model using OpenFOAM. The `simpleFuelCell` model has been deliberately kept basic in order to introduce simple electrochemical concepts, such as the stoichiometric ratio and reaction order, in comparison to subsequent chapters of this book, where much more complex algorithms are developed. The `simpleFuelCell` case is readily built-up from the base case for incompressible laminar flow of Newtonian fluids (`icoFoam`), by adding terms for the electrochemical reaction and species continuity (mass transfer). Full details are provided. The model may also be run bypassing the solution for the continuity and pressure-corrected momentum equations by fixing the stream-wise velocity to a constant value. The results are compared to an exact analytical solution and show near perfect agreement. The case can be used in-class, for university teaching purposes, e.g., in a first course on electrochemical modelling. If desired, the user may subsequently relax the model assumptions to investigate the impact on model accuracy or computational performance.

**Keywords** Fuel cell · Electrolyser · Computational fluid dynamics · Electrochemistry

The original version of this chapter was revised: The ESM material has been updated. The correction to this chapter is available at https://doi.org/10.1007/978-3-030-92178-1_10

**Supplementary Information** The online version contains supplementary material available at https://doi.org/10.1007/978-3-030-92178-1_1.

S. Beale (✉)
Institute of Energy and Climate Research, Forschungszentrum Jülich GmbH, IEK-14, 52425 Jülich, Germany
e-mail: s.beale@fz-juelich.de

Mechanical and Materials Engineering, Queen's University, Kingston, ON K7L 3N6, Canada

© Springer Nature Switzerland AG 2022, corrected publication 2022

S. Beale and W. Lehnert (eds.), *Electrochemical Cell Calculations with OpenFOAM*, Lecture Notes in Energy 42, https://doi.org/10.1007/978-3-030-92178-1_1

## Nomenclature

| | |
|---|---|
| $A$ | inverse Tafel slope |
| $c$ | concentration, kg/m$^3$ |
| $D$ | diffusivity, m$^2$/s |
| $E$ | open circuit voltage, V |
| F | Faraday's constant, $96485 \times 10^3$ C/kmol |
| $H$ | height, m |
| $i$ | Current, A |
| $j$ | current density, A/m$^2$ |
| $k_R$ | reaction rate constant, m/s |
| $L$ | length, m |
| M | molecular weight, molar mass, kg/kmol |
| $\dot{m}''$ | mass flux, kg/(m$^2$·s) |
| $N_{\text{iter}}$ | iteration number |
| $n$ | normal direction |
| $p$ | pressure, partial pressure, N/m$^2$ |
| $p_\alpha$ | partial pressure, N/m$^2$ |
| R | universal gas constant, 8.31446 J/(mol·K) |
| $R$ | residual [units of solved variable], resistance Ohm/m$^2$ |
| $\boldsymbol{S}$ | surface area vector, m$^2$ |
| $S$ | surface area magnitude, m$^2$ |
| $T$ | temperature, K |
| $\boldsymbol{U}$ | velocity, m/s |
| $U$ | stream-wise velocity, m/s |
| $V$ | cell voltage, V |
| $V_W$ | cross-wise velocity, wall velocity, m/s |
| $x$ | displacement, m |
| $X$ | mole fraction |
| $y$ | displacement, m |
| $Y$ | mass fraction |
| $Z$ | charge number |
| $z$ | displacement, m |

## Greek letters

| | |
|---|---|
| $\alpha_R$ | relaxation parameter |
| $\alpha_w$ | net protonic drag coefficient |
| $\gamma$ | reaction order |
| $\delta$ | cell half-width, m |
| $\lambda$ | stoichiometric coefficient |
| $\nu$ | kinematic viscosity, m$^2$/s |

$\rho$ density, $kg/m^3$
$\phi$ general scalar variable [any units]

## Subscripts

$H_2$ hydrogen
$H_2O$ water
$i$ inlet
$O_2$ oxygen
$o$ outlet
$P$ nodal value
$R$ reaction
$t$ transformed substance state
$W$ wall
0 ambient, external, reference value

# 1 Introduction

In recent years, a number of models of electrochemical devices, such as fuel cells and electrolysers, have been developed by engineers and scientists. These range, in complexity, from simple 0-D and 1-D analyses which run on a personal computer, to detailed transient 3-D cell and stack models, often requiring the use of supercomputers employing thousands of compute cores to obtain solutions. In recent years, the availability of well-written open source computational fluid dynamics (CFD) codes such as OpenFOAM facilitate the construction of multi-physics multi-scale models which are designed and developed from a 'top down' approach. One example of this is the `openFuelCell` code (Beale et al. 2016). In this chapter the reader is introduced to the subject of modelling electrochemical devices from a 'bottom up' approach. It is shown how to build the simplest possible 2-D electrochemical model, namely a half-cell, using OpenFOAM. Computational modelling generally requires that model closures be made. Details are provided on the specific model equations employed and the salient portions of the source code are provided with explanations. This raises the important topic of validation and verification:

> "Following Boehm (1981) and Blottner (1990), we adopt the succinct description of verification as "solving the equations right" and of validation as "solving the right equations."
>
> Patrick. J. Roache (1998).

This chapter is therefore concerned with verification of the cell model, by means of comparison of the results of the model calculations with a known analytical solution over a range of conditions. Thus, the reader is introduced to a number

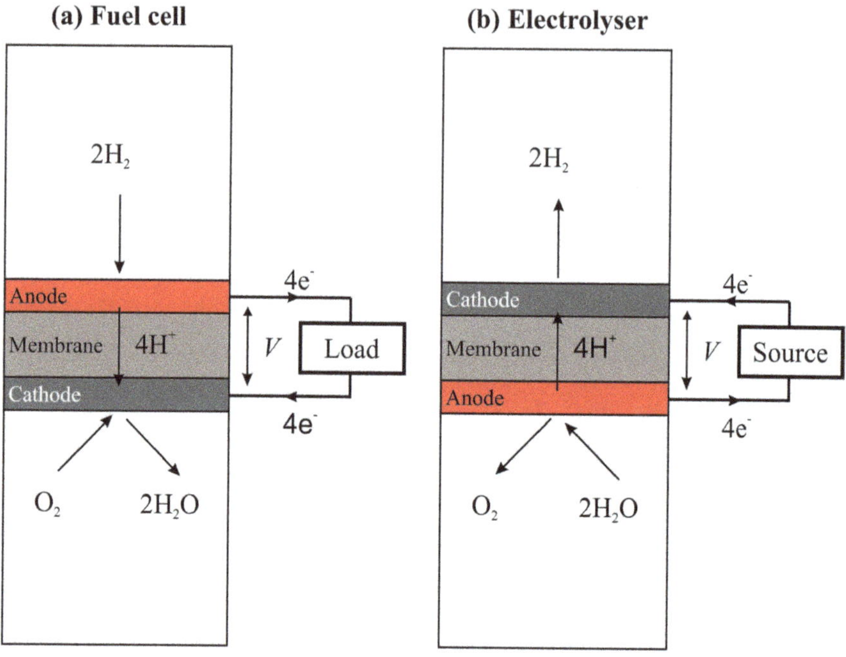


**Fig. 1** Schematic of an idealised polymer electrolyte cell: **a** fuel cell mode; **b** electrolyser mode

of concepts and definitions specific to electrochemistry, from an introductory position. Following this introduction, the reader should find him/herself in a position to construct electrochemical models of increasing complexity.

At the heart of any electrochemical cell are two electrodes; an anode and a cathode, as well as an electrically-conducting electrolyte/membrane, as is illustrated schematically in Fig. 1. In a polymer electrolyte fuel cell (PEFC), as portrayed in Fig. 1a, two half-reactions take place as follows: Hydrogen is oxidised at the anode with electrons passing through the load and protons crossing through the membrane. At the cathode, the protons combine with oxygen and electrons to form product water, i.e.,

$$\text{Anode}\quad 2\text{H}_2 \rightarrow 4\text{H}^+ + 4\text{e}^- \tag{1}$$

$$\text{Cathode}\quad 4\text{H}^+ + 4\text{e}^- + \text{O}_2 \rightarrow 2\text{H}_2\text{O} \tag{2}$$

The overall mass flux, $\dot{m}''$, is just that of the water produced less the oxygen consumed. A hydrogen fuel cell operated in reverse is simply a water electrolyser, also known as a polymer electrolyte electrolyser cell (PEEC). In that case, the water is converted into feedstock hydrogen and oxygen, as is shown in Fig. 1b. Under these circumstances, water is split into oxygen, protons, and electrons at the anode, and the hydrogen protons are simultaneously reduced (combined with the electrons) to form hydrogen at the cathode.

$$\text{Cathode}\quad 4H^+ + 4e^- \rightarrow 2H_2 \tag{3}$$

$$\text{Anode}\quad 2H_2O \rightarrow 4H^+ + 4e^- + O_2 \tag{4}$$

To avoid confusion; in this work the electrodes will be referred to as the oxygen electrode, and the hydrogen electrode. To date, there has been substantially more interest in the modelling of fuel cells than electrolysers; however, this situation is changing somewhat. The `simpleFuelCell` code, described in this chapter, was originally developed as a fuel cell model. It was subsequently adapted to also consider water electrolysis.

A number of expressions have been derived to describe the reaction kinetics. Among the simplest of these is the Tafel equation[1] (Gileadi 1993), which may be written, for instance for the oxygen electrode, as:

$$j = j_0 \left( \frac{Y}{Y_{ref}} \right)^{\gamma} \exp(A\eta) \tag{5}$$

where $j$ is current density (A/m$^2$), and $\eta$ is the so-called activation overpotential (Bockris et al. 2000), $Y = Y_{O_2}$ is the mass fraction of oxygen for a fuel cell, whereas $Y = Y_{H_2O}$ for an electrolyser, and $Y_{ref}$ is a reference value of $Y$. This can be, but is not necessarily, equal to the inlet value, $Y_i$. The term $j_0$ is the exchange current density and is generally given by an Arrhenius-type expression as a function of temperature. Equation (5) is not valid for $j/j_0 \leq 1$. The Tafel equation may be considered an expression for the additional potential or 'overpotential', $\eta$, (in units of volts) required to transfer charge across an electrode, in other words the deviation in the cell potential, $V$, from the ideal or open-circuit potential, $E$, when $j = 0$,

$$V = E \mp \eta \tag{6}$$

associated with the finite-rate reaction. For an electrolyser, $V > E$, and the overpotential $\eta = V - E$, is an expression of the fact that more electrical energy must be supplied than can be removed as chemical energy, at a finite-rate. Conversely, for a fuel cell, while $\eta$ is still referred-to universally as the 'overpotential', the cell voltage, $V$, is in fact actually less than the ideal potential, $E > V$: The reader will note that in practical electrochemical cells, additional voltage losses also result from other causes, such as mass transport and ohmic resistance. The constant, $A$, in Eq. (5) is sometimes called the inverse Tafel slope, due to the common practice of plotting $\eta$ versus $\log j$, referred to by electrochemists as a 'Tafel plot'.

Generally speaking, the hydrogen reaction is much faster than the oxygen one and so the former can be completely neglected, at least for a first order analysis, i.e.,

[1] After the Swiss chemist, Julius Tafel (1862–1918).

$\eta = \eta_{O_2}$ and $\eta_{H_2} = 0$ in Eq. (6). The oxygen mass flux, $\dot{m}''_{O_2}$, at the oxygen electrode is related to the current density, $j$, according to Faraday's law(s) of electrolysis:

$$\dot{m}''_{O_2} = \mp \frac{M_{O_2} j}{Z_{O_2} F} \tag{7}$$

where $M_{O_2} = 32$ kg/kmol is the molar mass of oxygen and $Z_{O_2} = 4$ is the charge number (valence) of the oxygen molecule. The negative sign is for fuel cell mode, which indicates that reactant oxygen is consumed by the reaction (employing the convention that $\dot{m}''$ is positive for injection and negative for suction). Conversely, the positive sign is for electrolyser mode and indicates that oxygen is being produced, i.e., $\dot{m}'' > 0$ applies to a fuel cell and $\dot{m}'' < 0$ to an electrolyser. The reader will note that this convention is commonly employed in texts on heat and mass transfer (heat added is positive, heat removed is negative). This is precisely the opposite to the convention adopted in OpenFOAM. Similarly, water is generated or removed such that:

$$\dot{m}''_{H_2O} = \pm \frac{M_{H_2O} j}{Z_{H_2O} F} \tag{8}$$

where $M_{H_2O} = 18$ kg/kmol and $Z_{H_2O} = 2$. Thus, in a fuel cell, even though mass in the form of water is being produced at the oxygen electrode, oxygen is consumed. In an electrolyser oxygen is being produced, and water consumed. The total mass flux, $\dot{m}''$, is:

$$\dot{m}'' = \rho V_W = \dot{m}''_{H_2} = \dot{m}''_{H_2O} - \dot{m}''_{O_2} \tag{9}$$

## 2 One-Dimensional Electrochemical Cell Analysis

Kulikovsky et al. (2004, 2005) derived an analytical solution for a one-dimensional (1-D) idealisation of a PEFC as a single oxygen electrode. The basic hypothesis was that mass transfer can, like heat transfer, be treated passively; in other words, the mass flux at the wall does not lead to substantial changes in the mixture density and velocity. This implies no overall change in mass continuity, but only in species concentration. The additional assumption of perfect mixing means that the individual species velocities in the stream-wise direction are all equal to the bulk value. The model assumptions are listed in detail below, followed by the mathematical expression for mass fraction, $Y$, and current density, $j$. The reader interested in the derivation of the analytical solution is referred to Appendix I, where it is given as a conventional mechanical engineering/control-volume analysis based on a mass-based formulation, rather than the molar form, as originally derived by Kulikovsky et al.

## 2.1 Model Assumptions

- The hydrogen oxygenation/reduction reaction is very 'fast' in comparison to the oxygen reduction/evolution, i.e., the anodic activation overpotential is negligibly small.
- The mixture density, $\rho$, is assumed to be constant.
- The oxygen-side streamwise velocity, $U$, is constant and the crosswise velocity is zero, in the equations for conservation of mass and momentum, but non-zero for species conservation.
- There is perfect mixing in the oxidant-side crosswise direction, and conversely no streamwise (axial) diffusion, i.e., only bulk convection.
- The open-circuit potential, $E$, Eq. (6), is constant, while the ohmic resistances of all internal components are negligibly small.
- The oxygen reaction-order, $\gamma$, the inverse-Tafel slope, $A$, in Eq. (5) and the cathodic activation overpotential, $\eta$, are all constant, and so the local current density is a function of the oxygen mass fraction, $Y$, and reaction order, $\gamma$, only.
- The temperature is constant throughout the cell.
- Water transport in the membrane is not considered.
- There is no crossover or leakage of gas from one electrode to the other
- The solubility of oxygen in liquid at the reaction sites is not considered an issue.

Based on the these assumptions, it is possible to derive the analytical solution, presented below.

# 3 Analytical Solution

The equation to be solved is:

$$\frac{\rho U}{M_\alpha}\frac{dY_\alpha}{dx} = -\frac{j}{Z_\alpha H \mathrm{F}} \tag{10}$$

where $j$ is given by Eq. (5), and $H$ is the height of the channel, see Fig. 2. Since this is a 2-D formulation, the width, $W$ may conveniently be chosen as unity, $W = 1$. The subscript $\alpha$ was used in Appendix I to distinguish between the different species; for a fuel cell $\alpha = O_2$ whereas for an electrolyser $\alpha = H_2O$. With this clearly understood, the subscript will be discarded henceforth, and the mass fraction denoted simply by $Y$. The solution for the mass fraction, $Y(x)$, is as follows:

$$\frac{Y}{Y_i} = \begin{cases} \left\{1 - \left[1 - \left(1 - 1/\lambda\right)^{1-\gamma}\right]\frac{x}{L}\right\}^{1/(1-\gamma)} & \gamma \neq 1 \\ \left(1 - 1/\lambda\right)^{x/L} & \gamma = 1 \end{cases} \tag{11}$$

where $Y_i$ is the inlet mass fraction of oxygen. The current density, $j(x)$, is given by:

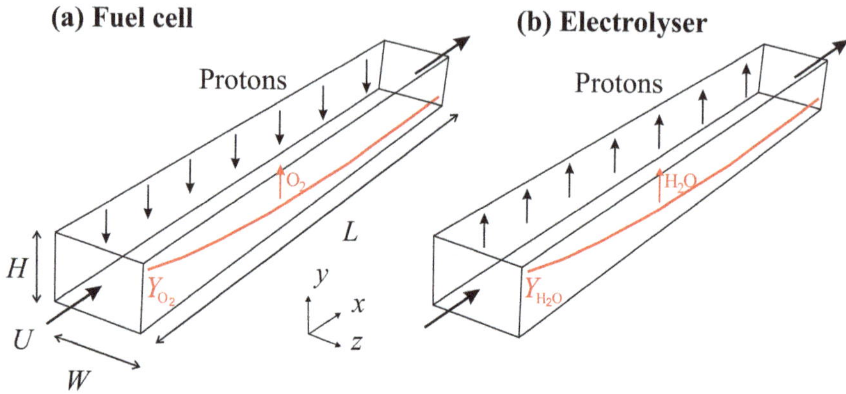


**Fig. 2** Idealised electrochemical cell showing the geometry, notation, and boundary conditions. Not to scale. **a** Fuel cell mode, the oxygen mass fraction decreases as oxygen is consumed by the reaction even though net mass, in the form of water, is being added: **b** Electrolyser mode; the water mass fraction decreases in the main flow direction as the feedstock water is consumed by the electrochemical reaction and oxygen is produced

$$\frac{j}{\overline{j}} = \begin{cases} \lambda\left(\frac{1-(1-1/\lambda)^{1-\gamma}}{1-\gamma}\right)\left\{1-\left[1-(1-1/\lambda)^{1-\gamma}\right]x/L\right\}^{\gamma/(1-\gamma)} & \gamma \neq 1 \\ -\lambda\left[\ln(1-1/\lambda)\right](1-1/\lambda)^{x/L} & \gamma = 1 \end{cases} \quad (12)$$

where $\overline{j}$ is the mean current density.

$$\overline{j} = \frac{1}{L}\int_0^L j dx \quad (13)$$

The stoichiometric ratio, λ, is defined for a fuel cell as:

$$\lambda = \frac{\text{maximum current}}{\text{actual current}} = \frac{i_{\max}}{i} \quad (14)$$

The maximum current corresponds to 100% of the oxygen/water at the inlet being consumed. NB: The actual current is just $i = \overline{j}l$, for the 2-D problem considered here, assuming $W = 1$. For constant ρ, $\boldsymbol{U}$:

$$\lambda = \frac{\overline{Y}_i}{\overline{Y}_i - \overline{Y}_o} \quad (15)$$

where $\overline{Y}_i$ and $\overline{Y}_0$ are the mass-averaged inlet and outlet mass fractions of oxygen for a fuel cell and water for an electrolyser. Clearly, $\lambda = 1$ implies that all of the oxygen/water is consumed, whereas $\lambda = \infty$ implies an open circuit, $i = 0$, no reaction. It follows that $1 \leq \lambda \leq \infty$. The analytical solution, below, was originally developed to model a PEFC, however, it may also be applied to PEECs where the mass flow (of hydrogen protons), is reversed, as is shown in Fig. 2b, and oxygen is produced with $\overline{Y}_i < \overline{Y}_o$. For an electrolyser, the oxygen stoichiometric number would be negative, however, a water stoichiometric can be, and is, similarly defined. The reader will note that the quantity $1/\lambda$ is sometimes referred to as the 'utilisation', namely the fraction of the total oxygen/water (or hydrogen on the hydrogen electrode side) that is utilised. Thus, $1 - 1/\lambda$ is that fraction of the total not utilised in the reaction; the term 'non-utilisation' could perhaps be coined.

## 4 Model Equations

The basic equations solved by `icoFoam` (Greenshields 2018) are continuity and momentum in the incompressible form.

$$\nabla \cdot \boldsymbol{U} = 0 \tag{16}$$

$$\nabla \cdot (\boldsymbol{U}\boldsymbol{U}) = -\nabla(p/\rho) + \nabla \cdot (\nu \operatorname{grad} \boldsymbol{U}) \tag{17}$$

These must be modified by the addition of a mass fraction equation.

$$\nabla \cdot (\boldsymbol{U}Y) = \nabla \cdot (\boldsymbol{D} \operatorname{grad} Y) \tag{18}$$

The solution of Kulikowsy et al. (2004, 2005) is based on the assumption that mixing is 'perfect', i.e., no cross-wise diffusion gradient, but conversely there exists a finite stream-wise gradient. One possible formulation is to treat the diffusivity, $\boldsymbol{D}$, as being directional, in nature. OpenFOAM admits that the diffusivity, $\boldsymbol{D}$, may be prescribed as a second order tensor, for example;

$$\boldsymbol{D} = \begin{pmatrix} D_{11} & D_{12} & D_{13} \\ D_{21} & D_{22} & D_{23} \\ D_{31} & D_{32} & D_{33} \end{pmatrix} = \begin{pmatrix} 10^{-15} & 10^{-15} & 10^{-15} \\ 10^{-15} & 1 & 10^{-15} \\ 10^{-15} & 10^{-15} & 1 \end{pmatrix} \tag{19}$$

which ensures diffusion dominates the $y$ and $z$ cross-wise directions, and convection the stream-wise $x$ direction. The main flow (stream-wise) is in the $x$-direction, with mass diffusion (cross-wise) in the $y$-direction as shown in Fig. 3. The $z$-direction is redundant (2-D problem).

The wall boundary condition may be written as:

$$\left.\frac{\partial Y}{\partial n}\right|_W = \frac{V_W}{D}(Y_T - Y_W) \tag{20}$$

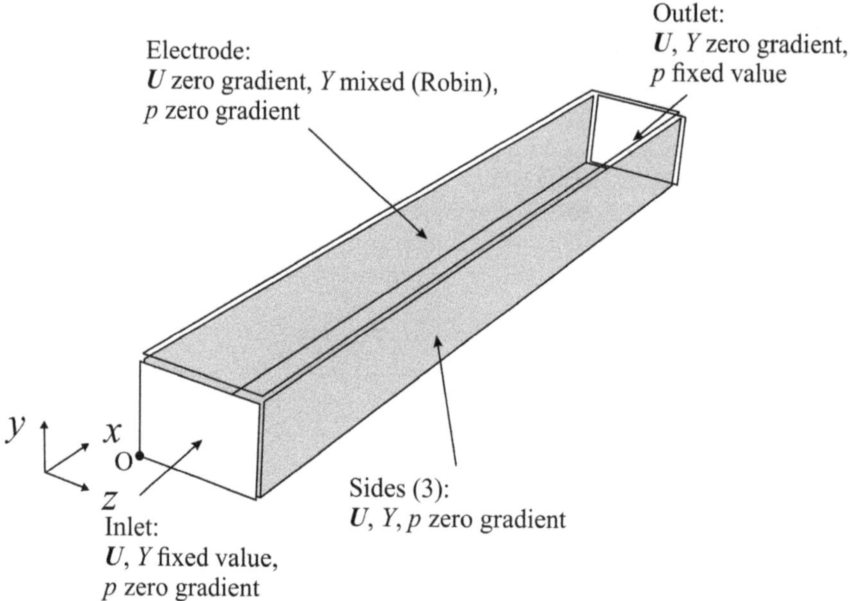


**Fig. 3** Boundary patches and values employed, for the case when PISO is employed with wall slip. The gas mixture flows in the $x$-direction with the electrode located in the plane $y = y_{max}$. The 3 sides (shaded) are located at $z = z_{min}$, $z_{max}$, and $y = y_{min}$, respectively. Since the problem considered is 2-D, the $z$-direction is redundant. Not to scale

where it is understood that $D = D_{22}$ in Eq. (19), $V_W$ is the wall mixture velocity, $Y_w$ is the wall value of $Y$, and $Y_T$ the so-called *transferred substance state* (T-state) value[2] (Spalding 1960, 1963; Kays et al. 2005; Beale 2004, 2005, 2007, 2015). The T-state value for species α is just the ratio of the mass flux of the individual species α, to the total mass flux.

$$Y_T = \frac{\dot{m}''_\alpha}{\dot{m}''} \tag{21}$$

For a non-reacting process, such as conventional membrane filtration $0 \leq Y_T \leq 1$, however for chemical and electrochemical reactions $Y_T$ is not bounded, $-\infty \leq Y_T \leq +\infty$. The total mass flow here is just $\dot{m}'' = \dot{m}''_{H_2} = \dot{m}''_{H_2O} - \dot{m}''_{O_2}$. Table 1 gives the T-state values for $H_2$, $O_2$, and $H_2O$. It should be noted that these values assume no net water transport, i.e., water is only produced/consumed by the reaction and not by pressure-gradient-driven convection, mass transport, or osmotic drag in the hydrated membrane. For the latter case, it can be shown (Beale 2015) that, e.g., for a fuel cell $Y_T = -8/(1 + 18\alpha_w)$, where $\alpha_w$ is the so-called net protonic drag coefficient. The

[2] Strictly speaking, $M_{O_2} = 31.998$ and $M_{H_2} = 2.01568$. So e.g. for $O_2$: $y_T = -M_{O_2} z_{H_2} / M_{H_2} z_{O_2} = -7.9372$ but, as is seen later, this does not affect the results in non-dimensional form.

**Table 1** Molar mass, charge number and transferred-substance state values for the reaction

| Electrode | Species, α | M | z | $Y_T$ |
|---|---|---|---|---|
| Oxygen | $O_2$ | 32 | 4 | -8 |
| Hydrogen | $H_2$ | 2 | 2 | 1 |
| Oxygen | $H_2O$ | 18 | 2 | +9 |

(hydrogen) wall velocity, $V_W$, is required for the species boundary condition, Eq. (20) only, not continuity and momentum Eqs. (16)–(17). This is obtained by combining Faraday's law of electrolysis, Eq. (7), with the Tafel equation, Eq. (5):

$$V_W = \left( \frac{\mathrm{M}_{\mathrm{H}_2} j_0 \exp(A\eta)}{\mathrm{Z}_{\mathrm{H}_2} \rho \mathrm{F} Y_{ref}} \right) Y^{\gamma} = k_R Y^{\gamma} \tag{22}$$

where $Y = Y_{\mathrm{O}_2}$ for a PEFC and $Y = Y_{\mathrm{H_2O}}$ for a PEEC. $k_R$ is the reaction rate constant defined as:

$$k_R = \pm \frac{\mathrm{M}_{\mathrm{H}_2} j_0 \exp(A\eta)}{\mathrm{Z}_{\mathrm{H}_2} \rho \mathrm{F} Y_{ref}} \tag{23}$$

which is a constant in the present model. The sign convention is somewhat arbitrary and has been chosen so that the direction of $V_W$ is in the negative coordinate direction (i.e., inward) for a fuel cell and positive (outward) for an electrolyser, in order to accommodate the fact that the hydrogen protons and electrons are being removed from the oxygen electrode in a PEEC, whereas they are being added in a PEFC (see Fig. 2).

NB: This formulation ensures that $V_W \to 0$ as $Y_W \to 0$, obviating the possibility of negative values of $Y$ from arising, i.e., the problem is bounded. Equation (20) is a Robin/mixed (linearised) boundary condition as is typically encountered in heat/mass transfer. OpenFOAM admits to such boundary condition prescriptions using a somewhat arcane procedure, as follows:

$$Y_W = fY_V + (1 - f)(Y_P + g\delta) \tag{24}$$

where $Y_P$ is the value of $Y$ at node P, $f$ is a *valueFraction*, $Y_v$ is a *refValue*, $g$ is a *refGradient*, $\delta$ is cell-centre to boundary distance,[3] $\delta = |\mathbf{P} - \mathbf{W}|$. Equation (20) may be implemented *either* by using a 'Neumann-like' fixed gradient approach:

$$f = \frac{\mathrm{P}}{1 + \mathrm{P}} \tag{25}$$

[3] The reader will note $\delta$ is distance and not the reciprocal of distance, as may be encountered in some OpenFOAM documentation: In OpenFOAM, deltaCoeff was historically chosen as 1/distance because floating-point multiplication (by 1/distance) is computationally cheaper than division (by distance).

$$g = 0 \tag{26}$$

*or* by means of a 'Dirichlet-like' fixed value formulation:

$$f = 0 \tag{27}$$

$$g = \frac{\mathrm{P}}{1+\mathrm{P}}\frac{Y_T - Y_P}{\delta} \tag{28}$$

where P is a cell Péclet number.

$$\mathrm{P} = V_w \delta / D \tag{29}$$

In both cases,

$$Y_v = Y_T \tag{30}$$

The Neumann form may be written as:

$$\frac{\partial Y}{\partial n} = \frac{1}{D} k_R Y_W^{\gamma} (Y_T - Y_W) \tag{31}$$

The reader will note that for $\gamma = 1$, the membrane boundary condition is now quadratic in $Y$. This class of boundary value problem is found elsewhere in the process and chemical industries, for example in membrane separation processes (Pharoah et al. 2000) where an osmotic pressure difference, which is a non-linear function of the mass fraction, is the driving force for mass transport. The matter is discussed further in Beale et al. (2013, 2020).

The reader will note that $\lambda$ may be calculated, either from the reaction rate, as $\lambda_R$, or from the outlet values of $U$ and $Y$, as $\lambda_O$, together with the inlet values:

$$\lambda_R = \sum_{\text{inlet}} Y\boldsymbol{U} \cdot \boldsymbol{dS} \Big/ \sum_{\text{electrode}} k_R Y^{\gamma} (Y_T - Y) dS \tag{32}$$

$$\lambda_O = \sum_{\text{inlet}} Y\boldsymbol{U} \cdot \boldsymbol{dS} \Big/ \left( \sum_{\text{inlet}} Y\boldsymbol{U} \cdot \boldsymbol{dS} - \sum_{\text{outlet}} Y\boldsymbol{U} \cdot \boldsymbol{dS} \right) \tag{33}$$

where $dS = |\boldsymbol{dS}|$ is the surface area and $\lambda_O = \lambda_R$ if and only if: (i) the Reynolds number, or more precisely the Péclet number for mass fraction, is sufficient large that axial diffusion may be discounted; and (ii) numerical convergence has been achieved. Indeed, $\lambda_O = \lambda_R$ may be used as a criterion of convergence for the code in place of the more usual residual sums, see Fig. 10, below.

For the purpose of generating a polarisation (current density vs. voltage) characteristic, fuel cells are not generally tested at constant inlet velocity; rather, above a certain minimum threshold, the inlet air/fuel in a fuel cell (oxidant/feedstock in an electrolyser) velocities are adjusted to maintain constant stoichiometric numbers, λ, as defined by Eq. (14) at both the electrode(s). This may be expedited in the numerical code by adjusting the next value of $U$ according to:

$$U = \left(1 + \alpha_R\left(\lambda^*/\lambda - 1\right)\right) \times U^* \tag{34}$$

where $\lambda^*$ and $U^*$ are values of the stoichiometric number and inlet velocity from the previous iteration, λ is the desired value and $\alpha_R$ is a relaxation parameter. Alternatively, the inlet mass fraction, $Y$, may be adjusted. Although it is, in fact, not strictly necessary to iterate using Eq. (34), provided that one assumes the analytical solution to be correct; when some of the assumptions given above are relaxed, an iterative approach such as Eq. (34) is simple to implement.

## 5 Practical Implementation

The case structure is illustrated in Fig. 4. The directory `run` contains the usual files/dictionaries. A dictionary in `/constant`, `cellProperties` holds the cell property values. `blockMesh` is used to generate a rectilinear mesh. The directory `src` contains the following custom C ++ source files: `createFields.H`, `simpleFuelCell.C`, `TstateBC.H`, `lambda.H`, `j.H`, `Re.H`. These are presented and discussed in detail in Appendix II, below. The listings provided here are known to run with OpenFOAM v6 as available from the OpenFOAM Foundation Ltd. The `simpleFuelCell` source code is updated by the present writer on a periodic basis. The reader who wishes to obtain the latest version of the `simpleFuelCell` code should therefore write directly to this author for the latest copy.

## 6 Results

The results were generated using a mesh of size 100 × 1 × 1, which was used to tessellate a region of size 0.2 × 0.01 × 0.01 m$^3$. Figures 5 and 6 show the non-dimensional mass fraction, $Y/Y_i$, and current density, $j/\overline{j}$, as a function of non-dimensional distance, $x/L$, for $\gamma = 1$. The numerical results are compared to the analytical solutions given in Eqs. (11) and (12). It can be seen that the level of agreement is very good. These profiles are similar to those that appear in the book by Kulikovsky (2010). The reader will note that as λ approaches unity, both the current density and mass fraction exhibit a highly non-linear profile, whereas for $\lambda \geq 2$, a reasonably linear profile is observed. It can be seen that the `simpleFuelCell` results for current density, $j/\overline{j}$, are near identical to the analytical results, whereas those for mass fraction, $Y$, are very slightly lower in value, for $\lambda = 1.01 - 1.1$. It is worth noting that the values of λ in the range of $1.01 \leq \lambda \geq 1.1$ (indicating that almost all

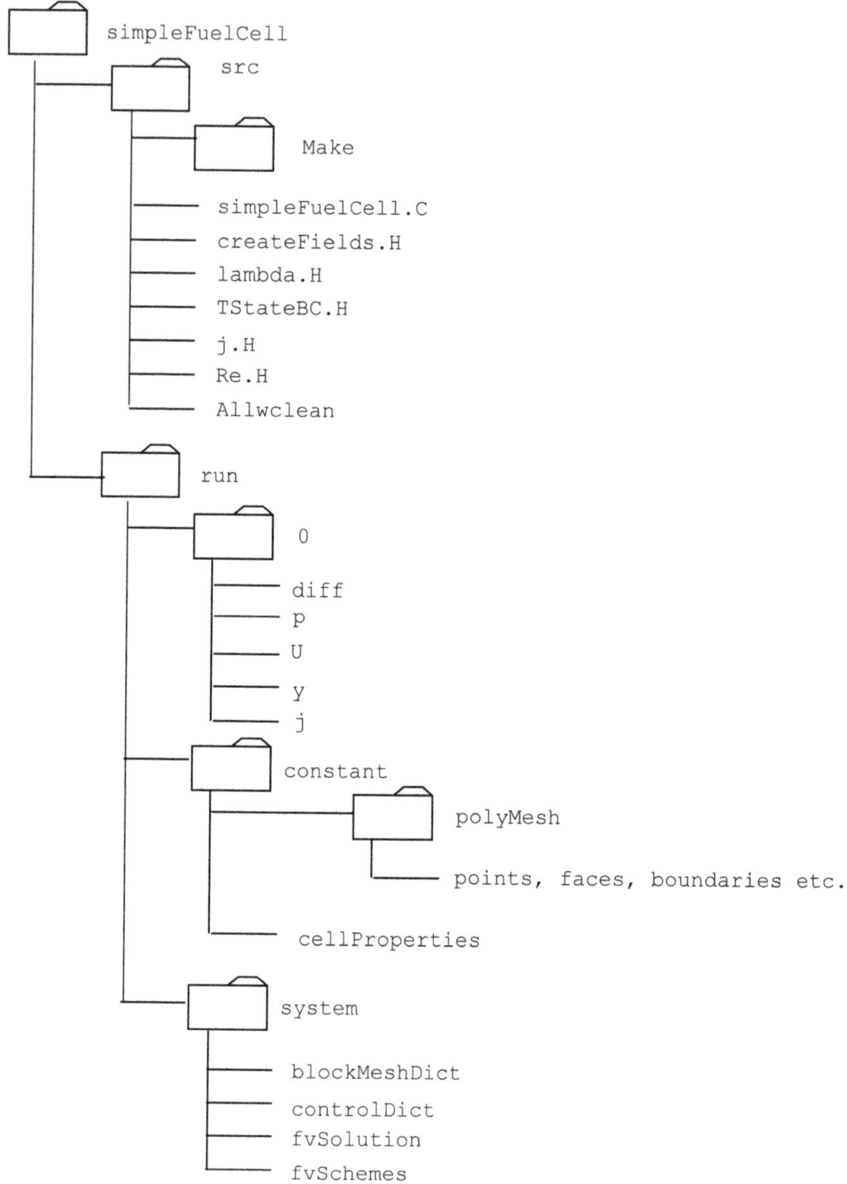


**Fig. 4** Simplified tree diagram for the files and directories in `simpleFuelCell`

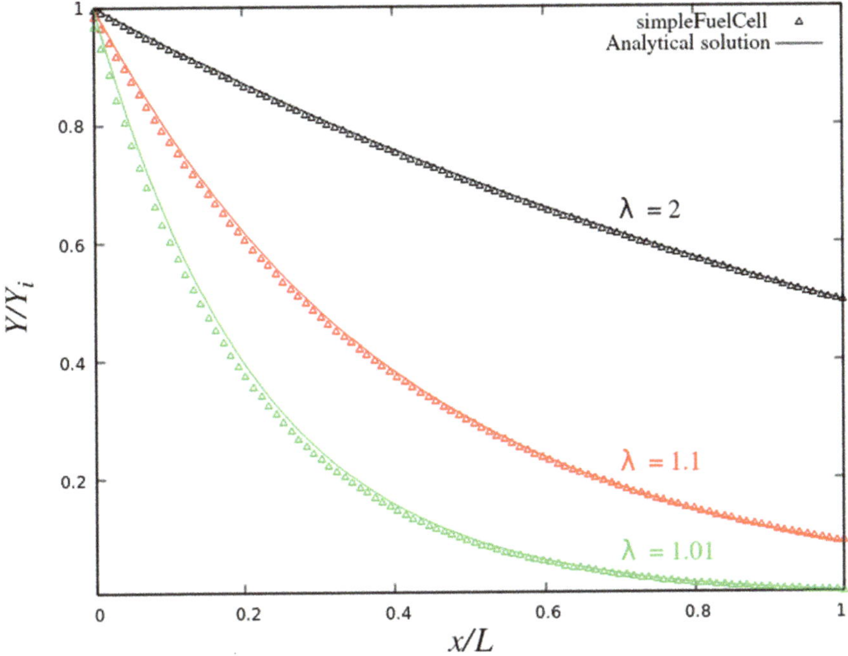


**Fig. 5** Normalised mass fraction $Y/Y_i$ versus non-dimensional distance, $\gamma = 1$

of the reactant is consumed) are not readily feasible in actual fuel cells/electrolysers; if they were, this would rapidly lead to cell degradation and destruction.

Figures 7 and 8 show $Y/Y_i$ and $j/\overline{j}$ versus $x/L$ for $\lambda = 2$, with the reaction order, $\gamma$, being varied. It can be seen that as the reaction order increases, so does the non-linearity of both the mass fraction and current density, with the reactant being consumed further upstream; the dependence on $Y^\gamma$, Eq. (22), becomes more pronounced as $\gamma$ increases.

Figures 9 and 10 show the convergence history for the performance calculations, with Fig. 9 showing the usual (sum of the normalised initial) residuals of $Y$. It can be seen that these decrease to a minimum threshold $R = 10^{-12}$, as set out in `run/system/fvSolution`, indicating that satisfactory convergence has been attained for the case $\lambda = 2$, $\gamma = 1$. It can be seen that the residuals decrease in a satisfactory manner over time, but it is observed that it is sometimes required to adjust the inlet $U$-value, iteratively, to obtain the set stoichiometric number, $\lambda$. This may be the cause of the convergence history not being a straight line on a logarithmic scale, as is seen in Fig. 9. Figure 10 is, perhaps, a better measure of convergence. It shows values of the stoichiometric parameter, $\lambda$, as a function of the iteration number, $N$, for the case $\lambda = 2$ and $\lambda = 1.1$, with $\gamma = 1$ in both cases. The values displayed are computed from both the electrochemical reaction rate, $\lambda_R$, Eq. (32), as well as the outlet flow rate, $\lambda_O$, Eq. (33). It can be seen that for $\lambda = 2$, $\lambda_R$ and $\lambda_O$, both converge within 200 iterations, whereas for $\lambda = 1.1$, while $\lambda_R$ converges

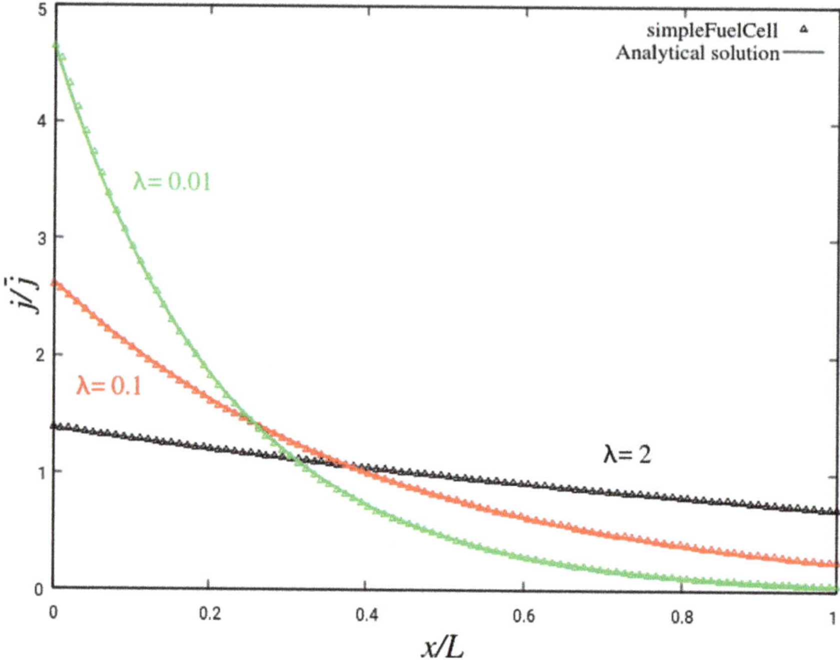


**Fig. 6** Current density normalized versus non-dimensional distance, $\gamma = 1$

very rapidly, $\lambda_O$ requires numerous iterations to asymptotically reach a stable value. While this history is clearly dependent upon the initial choice of $Y$-values and $U$-values (at $t = 0$) in `/0/Y` and `/0/U`, it is nonetheless true that near-wall values of $Y$ adjust rapidly to the reaction boundary condition, Eq. (20), whereas downstream conditions require time and the influence of convection to reach a stable condition. It is worth noting that the formulation of the chemical reaction, according to Eq. (20), is unconditionally stable, as $Y_w \to 0$, Eq. (22) ensures the wall velocity in Eq. (20), $V_W \to 0$ and thereby the boundary coefficient (which is an lvalue in the matrix of coefficients, rather than an rvalue or source term). This unconditional stability is not true for all mathematical formulations of the problem, as is discussed elsewhere (Beale et al. 2018, 2020).

## 7 Discussion

The results of Kulikovsky et al. (2004, 2005) are significant, as they demonstrate the important point that the two main parameters influencing PEFC performance are $\lambda$ and $\gamma$; a fact that was initially disputed, but later confirmed in physical experiments, and is at least approximately true for real electrochemical cells. It was shown here, in the present work, that the same results are equally applicable to PEECs. The only

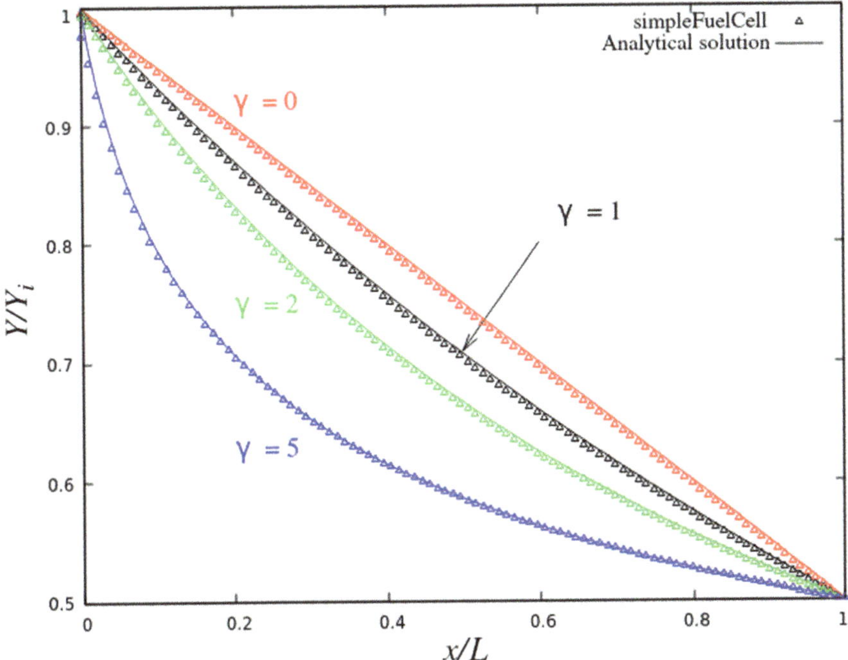


**Fig. 7** Normalised mass fraction $Y/Y_i$ versus non-dimensional distance for different reaction orders, $\lambda = 2$

changes to be made in the code are for the mass fraction boundary conditions: In a fuel cell $k_R$ is positive and $Y_T$ is negative ($Y_T$ = -8) whereas in an electrolyser $k_R$ is negative and $Y_T$ is positive ($Y_T = +9$), at least according the convention(s) adopted in this article.

As the operating conditions further depart from the idealised model assumptions, the actual mass fraction and current density will depart from those in Figs. 5 and 6. Nonetheless, the analytical solution can still be used to provide mathematical verification for more complex electrochemical cell calculations (Cao 2017; Zhang 2019; Zhang et al. 2018), especially at the debugging stage of code development.

The work of Kulikovsky (2010) is also remarkable in that it is a relatively rare example of an application outside the established disciplines of fluid mechanics and heat/mass transfer, where the concept of dimensional similitude is applied to sets of differential equations, which are solved in a non-dimensional form. In addition to the Reynolds number, it is also possible to define and compute other non-dimensional numbers, for example, a local Damköhler number, $\mathrm{Da} = k_R Y_W^{\gamma} H / D$, which gives the ratio of electrochemical reaction to diffusion terms[4] i.e., the 'speed' of the reaction. However, such non-dimensional numbers are seldom found in texts on electrochemical cells.

[4] For a 3-D geometry, the height, $H$, may be replaced with the hydraulic diameter, $D_h$.

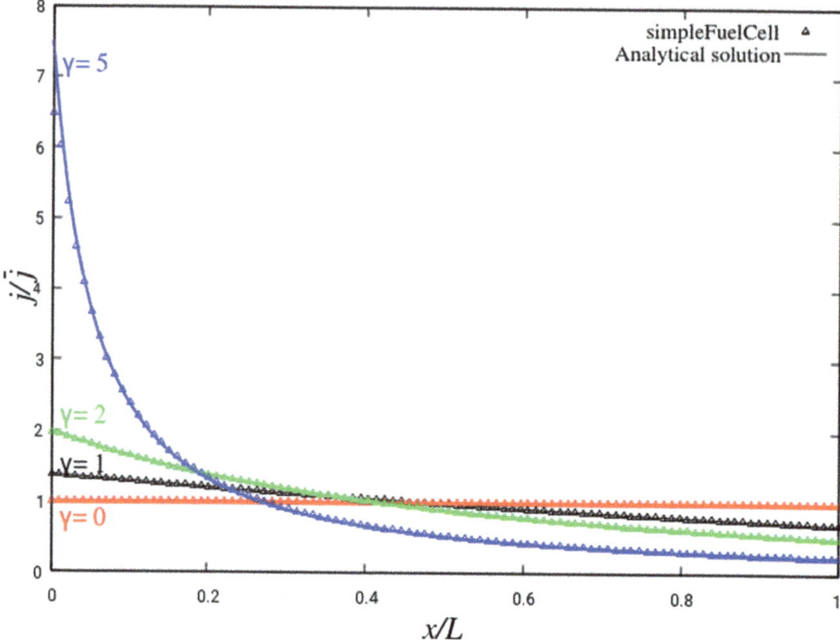


**Fig. 8** Current density normalized versus non-dimensional distance for $\lambda = 2$, with different reaction orders, $\gamma = 0,1,2,5$

## 8 Further Considerations

The base case for idealised electrochemical cell operation having been established, the user is free to adapt the model for more realistic conditions, and to investigate the impact of reducing the number of base assumptions. Some parameters to consider varying follow: these, by-and-large, correspond to relaxing the model assumptions given in Sect. 2.1, namely:

- The mixture density, $\rho$, need not be presumed constant, but rather a function of the make-up species. The ideal gas law for a mixture may be used to formulate $\rho$. At this point, it would be worth considering replacing the built-in `icoFoam` solver with another; although it is true to say that while the density of the fluid is not constant, due to the very large difference in the molar mass of $H_2$ and $H_2O$, the flow itself is still incompressible by any definition.
- The stream-wise velocity, $U$, need not be assumed to be a constant slug-flow, but rather a non-zero crosswise velocity based on Eq. (22), may be readily implemented as a boundary condition for the conservation equations for mass and momentum with the associated parabolic profiles in both the stream-wise and cross-wise velocities. This may be easily implemented by removing the following line in `/0/U`:

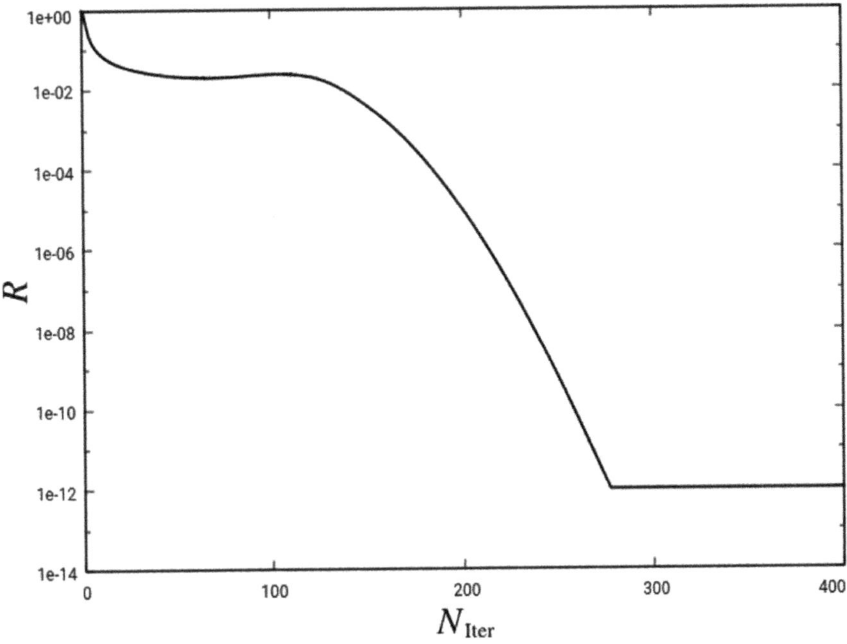


**Fig. 9** Normalised residual, $R$, for mass fraction, $Y$, versus iteration number, $N_{\text{Iter}}$. $\lambda = 2$, $\gamma = 1$. Fuel cell mode

```
membrane
   {
      type          zeroGradient;
   }
and by adding:
membrane
   {
      type          fixedValue;
      value         uniform (0 0 0);
   }
```

with a similar treatment accorded to the `wall` patch (the coding is there, but commented out). This will assure that: (a) the no-slip condition is imposed at the `wall`; and that (b) the membrane $U$ boundary condition is correctly set according to the identity for `U.boundaryField()[membranePatchID]` in `TstateBC.h`, as detailed in Appendix II. Of course, the user will have to re-run `blockMesh` with a reasonable number of cells and ensure that the Courant numbers are small enough to procure convergence.

The results for a zero order reaction $\gamma = 0$, may then be compared with the analytical solution of Berman (1953) for fluid flow, and the solution of Sherwood et al. (1963, 1965) for developing and fully-developed mass transfer; see, for example, Beale (2005), for an overview of those comparisons. A mathematical description of

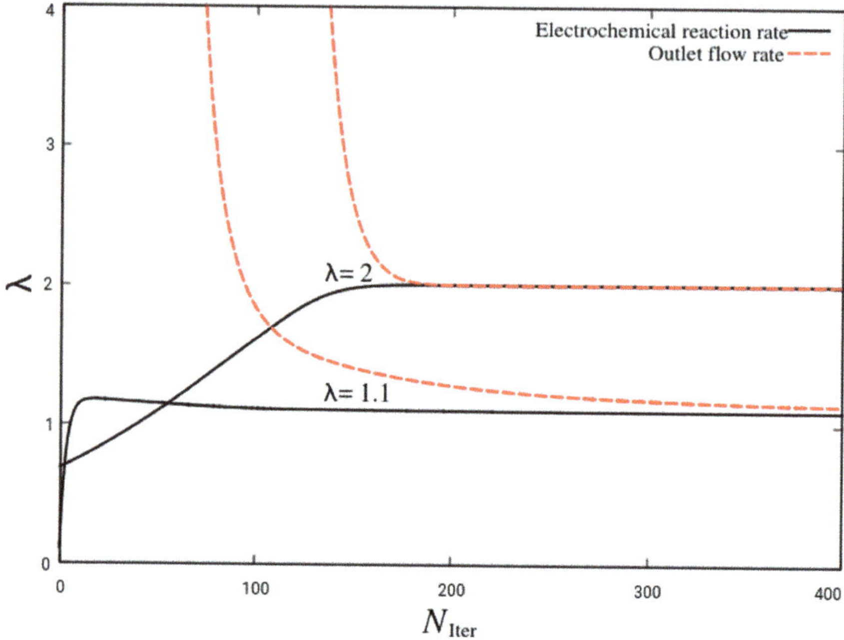


**Fig. 10** Convergence history for stoichiometric number, $\lambda$, versus iteration number, $N_{\text{Iter}}$, based on the electrochemical reaction rate, $\lambda_R$, and outlet flow rate, $\lambda_O$

the generalised wall boundary condition with injection/suction may be found in the recent chapter by Beale et al. (2020).

- Perfect mixing: The diffusion coefficient, $\boldsymbol{D}$, Eq. (19), may be based on more realistic values, rather than the artificial assumption of perfect mixing. N.B.: This once again will be a function of $Y$ and will, in general, be non-homogeneous due to the presence of porous transport layers as well as open channels in electrochemical cells, i.e., 'effective' values. This is an example of volume-averaging, also referred to nowadays as (*sic*) 'homogenisation'.
- The activation overpotental, $\eta$, need not be assumed to be constant. In this chapter/code, a single value for the reaction rate, $k_R$, was set in `constant/cellProperties` to describe the reaction kinetics, rather than prescribing the individual values of $\eta$, $j_0$, $A$, $Y_{ref}$. This conveniently renders the formulation in a form familiar to the mechanical/chemical engineer; however, an electrochemist would most certainly prescribe each of these four terms individually. From Eq. (23), the rate constant is $k_R \propto \exp(A\eta)$ and, from Eq. (6), $\eta = \pm(E - V)$. In reality, the open-circuit potential, $E$, Eq. (6), is not constant. From the laws of thermodynamics, one may deduce the change in $E$, according to a Nernst equation (Bockris et al. 2000) which, for a single active electrode, may be written in the following simplified form:

$$E = E_0 - \frac{RT}{ZF} \ln \left( \frac{\left(Y_{H_2O}/Y_{H_2O,ref}\right)^{1/2}}{\left(Y_{O_2}/Y_{O_2,ref}\right)} \right) \quad (35)$$

at any given temperature and pressure, and thereby the effects of the variation of $\Delta E$ and $\eta$ upon cell performance be readily investigated. This entirely neglects the effects of hydrogen variation on the hydrogen electrode, which was not considered here.

- The ohmic resistances of the internal components, $R$, may be significant. This is addressed by replacing Eq. (6) with an expression of the form:

$$\eta = \pm(E - V) - Rj \quad (36)$$

Equation (36) combines Kirchhoff's second law with Ohm's law. This identity, together with the Tafel equation, Eq. (5), and the Nernst equation yields three equations in three unknowns, for $E$, $j$ and $\eta$, which is a closed system of equations, i.e., a well-posed problem.

- Heat sources and sinks, and heat transfer mechanisms should be considered, i.e., an energy equation added. These are important, as they affect property values and, indeed, large temperature gradients can lead to failure in real electrochemical cells for a number of reasons. Constant temperature, like constant flow distribution and current density, is desired but not always attained in any given prototype: Electrochemical reactions like other reactions, generally involve the production or consumption of heat.
- Water transport through the membrane and in the porous transport layer and channels is extremely important. It can, and must, be accounted-for in PEFC/PEECs.
- Since both liquid water, and dissolved oxygen and gaseous oxygen are present at the reaction sites, the oxygen at the reaction site should also be considered, for instance, by means of Henry's law. This is generally considered a 'rate-limiting' or controlling factor, for diffusive transport. N.B.: In electrochemical texts diffusive losses are often evaluated in terms of voltage losses, and referred to as the 'concentration overpotential' or alternatively 'concentration polarisation'.

The relaxation of some of these latter assumptions is far from trivial. They will be further addressed in other chapters in this book.

## 9 Conclusion

A simplified 1-D formulation was derived for a highly idealised electrochemical cell exhibiting the essential features of a typical modern electrochemical cell. Initially, a

PEFC was considered, the model later being expanded to a PEEC. Analytical verification (Kulikovsky 2004; Kulikovsky et al. 2005) demonstrated the fidelity of the simple model in terms of the current density, $j$, and mass fraction, $Y$. Much of the model was, in fact, derived from previous, more complex 3-D work (Beale et al. 2016), and the contribution of all of the co-authors of that work is acknowledged. The present goal was to simplify the problem so that salient electrochemical phenomena could be isolated, with unnecessary physical and geometrical complexity (Reimer et al. 2019) being eliminated for the purposes of a preliminary or 'bottom-up' introduction. In subsequent chapters of this book, these important details will be introduced and amplified in a 'top-down' fashion, so that by the end of the book, the reader will find him/herself capable of building complex multi-physics models of electrochemical devices/processes. It is hoped that such a combined bottom-up/top-down approach, together with the open source paradigm, will lead to the development of ever more realistic models which will, in turn, contribute to the technological revolution required for environmentally-friendly energy conversion devices, such as fuel cells and electrolysers, to augment and eventually supplant conventional heat engines based on the consumption of non-renewable fossil fuels. For this to happen, the open source paradigm is of necessity a requirement, and the generation of transparent re-usable code of ever increasing applicability, to be freely shared amongst multiple users, obviates duplication and the creation of research silos, and thereby works to the benefit of all.

**Acknowledgements** The author would like to thank Prof. Hrvoje Jasak, Prof. Werner Lehnert, and Dr. Uwe Reimer for reviewing this manuscript, and providing suggestions for improving the text and the code. Some of the source code here was based on material developed in the article by Beale et al. (2016) and the author cordially acknowledges the contributions of all of the co-authors of that work; Dr. HaeWon Choi, Prof. Jon Pharoah, Mr. Helmut Roth, Prof. Hrvoje Jasak, and Prof. DongHyup Jeon.

## Appendix A1: Derivation of Analytical Solution

In view of the assumptions made in the model, any of the following may be selected as state-variable in deriving the analytical expression for oxygen distribution: normalized concentration, $c/c_i$, mole fraction,[5] $X/X_i$, partial pressure, $p/p_i$, partial density, $\rho/\rho_i$, mass fraction and $Y/Y_i$. All of these variables will follow the same distribution when suitably normalised. For a binary system, any of these variables may be selected as the state-variable, although historically, mass-based formulations are frequently encountered for diffusion modelling in CFD codes, due to the fact that the mass-averaged velocity corresponds to the velocity solved-for in the Navier–Stokes equations. Here, diffusion is essentially neglected and the matter is not especially

[5] Here, the following shorthand is employed: $p = p_{O_2}, \rho = \rho_{O_2}$ i.e., the *partial* pressure and density, *not the mixture values*. This would avoid ambiguity with use of the subscript '$i$', which here denotes an inlet value (not the value for species $i$).

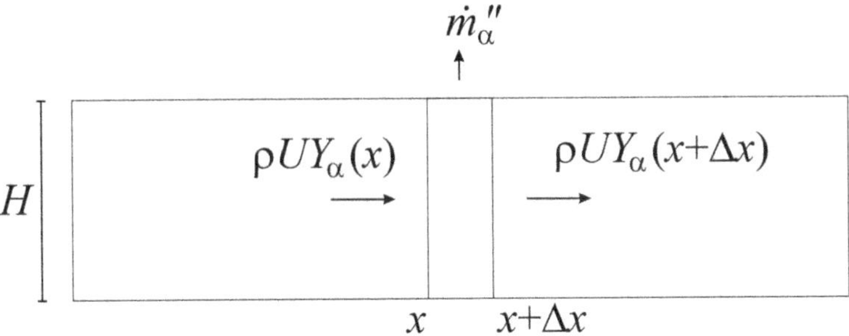


**Fig. A1.1.** Control volume analysis. For a fuel cell $\alpha = O_2$ whereas for an electrolyser $\alpha = H_2O$

important. The reader will note that if the wall velocity, $V_w$, was applied to continuity and momentum equations, then the mixture density, ρ, and mixture molar mass, M, would vary, as would the molar and mass-based distributions. Essentially, all solutions start with the following identity:

$$j = a\frac{d\phi}{dx} = b\phi^\gamma \tag{A1.1}$$

where $a$ and $b$ are constants and $\phi = c, X, p, \rho, Y$ as appropriate. Equation (A1.1) is a linear-first order equation that may readily be solved, first for $\phi$ and subsequently for $j$. For the convenience of the reader, a mass-based formulation based on a standard mechanical engineering control-volume formulation is derived with $\phi = Y$. Kulikovsky et al. considered instead, $\phi = c$.

For convenience, the symbol $Y_\alpha$ is used to denote the mass fraction of species α. For a fuel cell $Y_\alpha = Y_{O_2}$, whereas for an electrolyser $Y_\alpha = Y_{H_2O}$, the only difference between the two cases is that the net mass flux, $\dot{m}''$ (and $U$ and $j$) is/are inward (injection) for a fuel cell and outward (suction) for an electrolyser, however in both cases $\dot{m}''_\alpha$ is as shown since both oxygen/water are consumed by the reaction in a fuel cell/electrolyser at the oxygen electrode, i.e., removed from the control-volume. Without loss of generality it can be assumed that the width of the cell, Fig. 2, $W$ = 1 m. A control-volume for the problem at hand is illustrated in Fig. A1.1. If the velocity, $U$, and density, ρ, are presumed constant:

$$(\rho U Y_\alpha(x + \Delta x) - \rho U Y_\alpha(x))H = \dot{m}''_\alpha \Delta x \tag{A1.2}$$

$$\frac{dY_\alpha}{dx} = \frac{\dot{m}''_\alpha}{\rho U H} \tag{A1.3}$$

From Faraday's law of electrolysis, Eq. (7):

$$\dot{m}''_\alpha = -\frac{M_\alpha j}{Z_\alpha F} = \rho y_T k_R Y_\alpha^\gamma \tag{A1.4}$$

$$j = \left(\frac{Z_\alpha F \rho U H}{M_\alpha}\right)\frac{dY_\alpha}{dx} = \left(\frac{Z_\alpha F}{M_\alpha}\rho Y_T k_R\right) Y_\alpha^\gamma \tag{A1.5}$$

where $k_R$ is given in Eq. (23). Equation (A1.5) is one form of Eq. (A1.1), in mass units. Thus, Eq. (5) is obtained as:

$$\frac{dY_\alpha}{dx} = \frac{\rho Y_T k_R}{UH} Y_\alpha^\gamma \tag{A1.6}$$

Define $Y^* = Y_a / Y_{\alpha,i}$, where $Y_{\alpha,i} = Y_\alpha(0)$, $x^* = x/L$, so that

$$\frac{dY^*}{dx^*} = -(AY^*)^\gamma \tag{A1.7}$$

where

$$A = \left(\frac{M_{O_2} k_R}{2FHUY_{\alpha,i}}\right)\left(\frac{Y_{\alpha,i}}{Y_{\alpha,ref}}\right)^\gamma \tag{A1.8}$$

$$\begin{aligned} -\tfrac{1}{1-\gamma}(Y^*)^{1-\gamma} &= Ax^* + K \quad \gamma \neq 1 \\ -\ln(Y^*) &= Ax^* + K \quad \gamma = 1 \end{aligned} \tag{A1.9}$$

$y^* = 1$ at $x^* = 0$

$$K = \begin{cases} \frac{1}{1-\gamma} & \gamma \neq 1 \\ 0 & \gamma = 1 \end{cases} \tag{A1.10}$$

Thus,

$$Y^* = \begin{cases} (1-(1-\gamma)Ax^*)^{\frac{1}{1-\gamma}} & \gamma \neq 1 \\ \exp(-Ax^*) & \gamma = 1 \end{cases} \tag{A1.11}$$

Eliminate $A$, using Eq. (14):

$$\frac{Y_{\alpha,0}}{Y_{\alpha,i}} = 1 - \frac{1}{\lambda} = \begin{cases} (1-(1-\gamma)A)^{\frac{1}{1-\gamma}} & \gamma \neq 1 \\ \exp(-A) & \gamma = 1 \end{cases} \tag{A1.12}$$

$$A = \begin{cases} \left(1-(1-1/\lambda)^{1-\gamma}\right)/(1-\gamma) & \gamma \neq 1 \\ \exp(1-1/\lambda) & \gamma = 1 \end{cases} \tag{A1.13}$$

$$\boxed{\frac{Y_\alpha}{Y_{\alpha,i}} = \begin{cases} \left\{1-\left[1-(1-1/\lambda)^{1-\gamma}\right]x/L\right\}^{\gamma/(1-\gamma)} & \gamma \neq 1 \\ (1-1/\lambda)^{x/L} & \gamma = 1 \end{cases}} \tag{A1.14}$$

The current density is obtained by differentiation, as below:

$$j=-Z_\alpha FHU\frac{dY_\alpha}{dx}=\begin{cases} Z_\alpha FHUY_{\alpha,i}\left(\frac{1-(1-1/\lambda)^{1-\gamma}}{1-\gamma}\right)\left\{1-\left[1-(1-1/\lambda)^{1-\gamma}\right]x/L\right\}^{\gamma/(1-\gamma)} & \gamma\neq 1\\ Z_\alpha FHUY_{\alpha,i}\left[\ln\left(1-\frac{1}{\lambda}\right)\right]\left(1-\frac{1}{\lambda}\right)^{x/L} & \gamma=1\end{cases} \tag{A1.15}$$

$$j(x)=\left(\frac{Z_\alpha F\rho UH}{M_\alpha}\right)Y_\alpha(x)^\gamma \tag{A1.16}$$

$$j=\begin{cases}\left(\frac{Z_\alpha F\rho HU}{M_\alpha}\right)Y^\gamma_{\alpha,i}\left(\frac{1-(1-1/\lambda)^{1-\gamma}}{1-\gamma}\right)\left\{1-\left[1-(1-1/\lambda)^{1-\gamma}\right]x/L\right\}^{\gamma/(1-\gamma)} & \gamma\neq 1\\ \left(\frac{Z_\alpha F\rho HU}{M_{O_2}}\right)Y_{\alpha,i}\left[\ln\left(1-\frac{1}{\lambda}\right)\right]\left(1-\frac{1}{\lambda}\right)^{x/L} & \gamma=1\end{cases} \tag{A1.17}$$

The limiting or maximum current density is obtained for, $\lambda = 1$, as[6]:

$$j_{\max}=\left(\frac{Z_\alpha F}{M_\alpha}\rho y_T k_R\right)Y^\gamma_{\alpha,i} \tag{A1.18}$$

Thus,

$$\frac{j}{j_{\max}}=\begin{cases}\left(\frac{1-(1-1/\lambda)^{1-\gamma}}{1-\gamma}\right)\left\{1-\left[1-(1-1/\lambda)^{1-\gamma}\right]x/L\right\}^{\gamma/(1-\gamma)} & \gamma\neq 1\\ \left[\ln\left(1-\frac{1}{\lambda}\right)\right]\left(1-\frac{1}{\lambda}\right)^{x/L} & \gamma=1\end{cases} \tag{A1.19}$$

The average current is just,

$$\bar{j}=-\frac{1}{L}\int_0^L\frac{Z_\alpha F\rho Uh}{M_\alpha}\frac{dY_\alpha}{dx}dx=\frac{Z_\alpha F\rho Uh\left(Y_{\alpha,i}-Y_{\alpha,0}\right)}{M_\alpha L} \tag{A1.20}$$

So that:

$$\frac{\bar{j}}{j_{\max}}=\left(1-\frac{Y_{\alpha,o}}{Y_{\alpha,i}}\right)=\frac{1}{\lambda} \tag{A1.21}$$

$$\boxed{\frac{j}{\bar{j}}=\begin{cases}\lambda\left(\frac{1-(1-1/\lambda)^{1-\gamma}}{1-\gamma}\right)\left\{1-\left[1-(1-1/\lambda)^{1-\gamma}\right]x/L\right\}^{\gamma/(1-\gamma)} & \gamma\neq 1\\ -\lambda\left[\ln(1-1/\lambda)\right](1-1/\lambda)^{x/L} & \gamma=1\end{cases}} \tag{A1.22}$$

[6] This is only strictly true when the Péclet number, Pe > 1.

## Appendix A2: Source Code

### run directory

### /constant/cellProperties

```
/*--------------------------------*- C++ -*----------------------------------
*\
  =========                 |
  \\      /  F ield         | OpenFOAM: The Open Source CFD Toolbox
   \\    /   O peration     | Website:  https://openfoam.org
    \\  /    A nd           | Version:  6
     \\/     M anipulation  |
\*---------------------------------------------------------------------------
*/
FoamFile
{
    version     2.0;
    format      ascii;
    class       dictionary;
    location    "constant";
    object      cellProperties;
}
// * * * * * * * * * * * * * * * * * * * * * * * * * * * * * * * * * * * * *
//

nu                  nu [ 0 2 -1 0 0 0 0 ]   15.0e-6;
dHyd                dHyd [ 1 0 0 0 0 0 0]   1.e-02;
tStateValue         tState [ 0 0 0 0 0 0 0] -8;
relax               relax [0 0 0 0 0 0 0]   0.3;
lambda              lambda [0 0 0 0 0 0 0]  2.0;
gamma               gamma [0 0 0 0 0 0 0]   1.0;
kReact              kReact [1 1 0 0 0 0 0]  0.006;
constlambda         true;
lpiso               true;
wall                false;

//    ************************************************************************
//
```

[Listing of `cellProperties` dictionary]

This dictionary contains values of kinematic viscosity, ν, hydraulic diameter $D_h$ (used to calculate Reynold number), oxygen T-state value $y_T = 8$, Eq. (21), relaxation parameter, α., stoichoiometry, λ, (if prescribed), reaction order, γ, and rate constant $k_R$, In addition the following logicals are present: `constlambda true` effects iterative adjusting of the inlet velocity until the desired value for `lambda` is reached. `lpiso true` implies `icoFoam` will be employed, otherwise the velocity field is simply fixed (and may be adjusted iteratively if `constlambda=true`). `wall true` will result in active rather than passive mass transfer i.e. boundary conditions in the continuity (pressure) and momentum equations otherwise these are disabled corresponding to the analytical solution given above.

## src directory

### simpleFuelCell.C

This is a modified version of `icoFoam`. The main change is that the solution for the scalar mass fraction, $Y$, has been added to the code.

```
/*---------------------------------------------------------------------------*\
  =========                 |
  \\      /  F ield         | OpenFOAM: The Open Source CFD Toolbox
   \\    /   O peration     | Website:  https://openfoam.org
    \\  /    A nd           | Copyright (C) 2011-2018 OpenFOAM Foundation
     \\/     M anipulation  |
-------------------------------------------------------------------------------
License
    This file is part of OpenFOAM.

    OpenFOAM is free software: you can redistribute it and/or modify it
    under the terms of the GNU General Public License as published by
    the Free Software Foundation, either version 3 of the License, or
    (at your option) any later version.

    OpenFOAM is distributed in the hope that it will be useful, but WITHOUT
    ANY WARRANTY; without even the implied warranty of MERCHANTABILITY or
    FITNESS FOR A PARTICULAR PURPOSE.  See the GNU General Public License
    for more details.

    You should have received a copy of the GNU General Public License
    along with OpenFOAM.  If not, see <http://www.gnu.org/licenses/>.

Application
    simpleFuelCell

Description
    Simple single electrode kinetics solver, based on icoFOAM.

Adapted from icoFOAM
by
Steven Beale
Forschungszentrum Juelich GmbH
2017-2020

\*---------------------------------------------------------------------------*/

#include "fvCFD.H"
#include "pisoControl.H"
// >>>>>>>>>>>>>>>>>>>>>>>>>>>>>>>>>>>
#include "mixedFvPatchFields.H"// needed for T-state boundary condition
// <<<<<<<<<<<<<<<<<< <<<<<<<<<<<<<<<<
// * * * * * * * * * * * * * * * * * * * * * * * * * * * * * * * * * * * * //

int main(int argc, char *argv[])
{
    #include "setRootCase.H"
    #include "createTime.H"
    #include "createMesh.H"

    pisoControl piso(mesh);

    #include "createFields.H"
    #include "initContinuityErrs.H"

    // * * * * * * * * * * * * * * * * * * * * * * * * * * * * * * * * * * //
```

```
    Info<< "\nStarting time loop\n" << endl;

    for (runTime++; !runTime.end(); runTime++)
//    while (runTime.loop())
// Cut out PISO
    {
        Info<< "Time = " << runTime.timeName() << nl << endl;
        #include "CourantNo.H"
        if (lpiso)
{//
        // Momentum predictor

        fvVectorMatrix UEqn
        (
            fvm::ddt(U)
          + fvm::div(phi, U)
          - fvm::laplacian(nu, U)
        );

        if (piso.momentumPredictor())
        {
            solve(UEqn == -fvc::grad(p));
        }

        // --- PISO loop
        while (piso.correct())
        {
            volScalarField rAU(1.0/UEqn.A());

            volVectorField HbyA("HbyA", U);
            HbyA = rAU*UEqn.H();
            surfaceScalarField phiHbyA
            (
                "phiHbyA",
                (fvc::interpolate(HbyA) & mesh.Sf())
              + fvc::interpolate(rAU)*fvc::ddtCorr(U, phi)
            );

            adjustPhi(phiHbyA, U, p);

            // Non-orthogonal pressure corrector loop
            while (piso.correctNonOrthogonal())
            {
                // Pressure corrector

                fvScalarMatrix pEqn
                (
                    fvm::laplacian(rAU, p) == fvc::div(phiHbyA)
                );

                pEqn.setReference(pRefCell, pRefValue);

                pEqn.solve();
                if (piso.finalNonOrthogonalIter())
                {
                    phi = phiHbyA - pEqn.flux();
                }
            }

            #include "continuityErrs.H"

            U = HbyA - rAU*fvc::grad(p);
            U.correctBoundaryConditions();
        }
```

```
}//End cut out PISO
//Add these lines for mass fraction y
       fvScalarMatrix yEqn
        (
              fvm::ddt(y)
            + fvm::div(phi, y)
            - fvm::laplacian(diff, y)
        );
        #include "TStateBC.H"
        yEqn.solve();
        y.correctBoundaryConditions();
    Info << "min,mean,max(y" << "): "
         << Foam::gMin(y) <<" , "
         << Foam::gAverage(y) <<" , "
         << Foam::gMax(y) << endl;
        #include "lambda.H"
        #include "Re.H"
        #include "j.H"
        runTime.write();
        Info<< "ExecutionTime = " << runTime.elapsedCpuTime() << " s"
            << "  ClockTime = " << runTime.elapsedClockTime() << " s"
            << nl << endl;
    }

    Info<< "End\n" << endl;

    return 0;
}


// ************************************************************************* //
```

[Listing of simpleFuelCell.C]

```
// Creates initial fields for simpleFuelCell
// Written Steven Beale
// Forschungszentrum Juelich GmbH
// 2017-2020

Info<< "Reading cellProperties\n" << endl;

IOdictionary cellProperties
(
    IOobject
    (
        "cellProperties",
        runTime.constant(),
        mesh,
        IOobject::MUST_READ_IF_MODIFIED,
        IOobject::NO_WRITE
    )
);
```

Additionally the file mixedFvPatchFields.H, has been included. This facilitates the solution for the T-state boundary condition, Eq. (20). In addition, the following files were also included: TStateBC.H, stoichiometricFactor .H, ReynoldsNumber.H, idensity.H. These are discussed below.

**createFields.H**

This is where the fields are read from the dictionary `run/cellProperties`. Note that the diffusion coefficient, diff, is defined as a `volumeTensorField`.

```
    dimensionedScalar nu(cellProperties.lookup("nu"));
    dimensionedScalar dHyd(cellProperties.lookup("dHyd"));
    dimensionedScalar tState(cellProperties.lookup("tStateValue"));
    dimensionedScalar relax(cellProperties.lookup("relax"));
    dimensionedScalar lambda(cellProperties.lookup("lambda"));
    dimensionedScalar gamma(cellProperties.lookup("gamma"));
    dimensionedScalar kReact(cellProperties.lookup("kReact"));
    Switch constlambda(cellProperties.lookup("constlambda"));
    Switch lpiso(cellProperties.lookup("lpiso"));
    Switch wall(cellProperties.lookup("wall"));
    dimensionedScalar ilambda = lambda/mag(lambda.value());
            Info<< "   nu = " << nu.value()
                << nl;
            Info<< "   dHyd = " << dHyd.value()
                << nl;
            Info<< "   tStateValue = " << tState.value()
                << nl;
            Info<< "   relax = " << relax.value()
                << nl;
            Info<< "   ilambda = " << ilambda.value()
                << nl;
            Info<< "   lambda = " << lambda.value()
                << nl;
            Info<< "   gamma = " << gamma.value()
                << nl;
            Info<< "   kReact = " << kReact.value()
                << nl;
            Info<< "   constlambda = " << constlambda
                << nl;
            Info<< "   lpiso = " << lpiso
                << nl;
            Info<< endl;
/*
dimensionedScalar nu
(
    "nu",
    dimViscosity,
    cellProperties.lookup("nu")
);

*/

Info<< "Reading field diff\n" << endl;

volTensorField diff
(
    IOobject
    (
        "diff",
        runTime.timeName(),
        mesh,
        IOobject::MUST_READ,
        IOobject::AUTO_WRITE
    ),
    mesh
);
```

```
Info<< "Reading field p\n" << endl;
volScalarField p
(
    IOobject
    (
        "p",
        runTime.timeName(),
        mesh,
        IOobject::MUST_READ,
        IOobject::AUTO_WRITE
    ),
    mesh
);


Info<< "Reading field U\n" << endl;
volVectorField U
(
    IOobject
    (
        "U",
        runTime.timeName(),
        mesh,
        IOobject::MUST_READ,
        IOobject::AUTO_WRITE
    ),
    mesh
);

volScalarField y
(
    IOobject
    (
        "y",
        runTime.timeName(),
        mesh,
        IOobject::MUST_READ,
        IOobject::AUTO_WRITE
    ),
    mesh
);

volScalarField j
(
    IOobject
    (
        "j",
        runTime.timeName(),
        mesh,
        IOobject::MUST_READ,
        IOobject::AUTO_WRITE
    ),
    mesh
);

#include "createPhi.H"


label pRefCell = 0;
scalar pRefValue = 0.0;
setRefCell(p, mesh.solutionDict().subDict("PISO"), pRefCell, pRefValue);
mesh.setFluxRequired(p.name());
```

[Listing of `createFields.H`]

```
// Computes T-state boundary condition
// Written Steven Beale
// Forschungszentrum Juelich GmbH
// 2017-2020

    label membranePatchID = mesh.boundaryMesh().findPatchID("membrane");
     if (membranePatchID==-1)
    {
      Info << "Failed to find patch named membrane for T-state boundary
condition"
            <<endl;
    }
else
{
Info << nl << "Solving mass fraction boundary condition" << endl;
scalarField mflux(phi.boundaryField()[membranePatchID].size(),0);
scalarField Pecl(phi.boundaryField()[membranePatchID].size(),0);
scalarField gammaS(phi.boundaryField()[membranePatchID].size(),1.0);
volScalarField& Ys = y;
mixedFvPatchScalarField& YsBC =
refCast<mixedFvPatchScalarField>
(
Ys.boundaryFieldRef()[membranePatchID]
);
const fvPatch& myPatch = mesh.boundary()[membranePatchID];
scalarField Delta = myPatch.deltaCoeffs();
scalarField yB = Ys.boundaryField()[membranePatchID].patchInternalField();
mflux = kReact.value()*pow(yB,gamma.value());
Info << "min,mean,max(yB" << "): "
     << Foam::gMin(yB) <<" , "
     << Foam::gAverage(yB) <<" , "
     << Foam::gMax(yB) << endl;

    // 2. Compute the velocity/mass flux

   Pecl = mflux/(gammaS*Delta);//Compute cell Peclet number for density =
1.
   Info << "min,mean,max(Pe" << "): "
     << Foam::gMin(Pecl) <<" , "
     << Foam::gAverage(Pecl) <<" , "
     << Foam::gMax(Pecl) << endl;
   Info << "min,mean,max(mflux" << "): "
     << Foam::gMin(mflux) <<" , "
     << Foam::gAverage(mflux) <<" , "
     << Foam::gMax(mflux) << endl;

// 3. Set the interface velocity condition from the mass flux

    if (wall)
{
U.boundaryField()[membranePatchID] ==
(
-1.0*(mflux)
*(mesh.Sf().boundaryField()[membranePatchID])
/(mesh.magSf().boundaryField()[membranePatchID])
);
      Info << "min,mean,max(U.boundaryField" << "): "
            << Foam::gMin(U.boundaryField()[membranePatchID]) <<" , "
            << Foam::gAverage(U.boundaryField()[membranePatchID]) <<" , "
            << Foam::gMax(U.boundaryField()[membranePatchID]) << endl;
}
```

## TstateBC.H

```
// 4. Set the scalar mass fraction Robin boundary condition
// Prescribed value formulation
//    YsBC.refValue() = TStateValue;
//    YsBC.valueFraction() = Pecl/(1.0+Pecl);// f
//    YsBC.refGrad() = 0.0;// g
// Prescribed gradient formulation
    YsBC.refValue() = tState.value();
    YsBC.refGrad() = (YsBC.refValue()-YsBC)*Delta*Pecl/(1+Pecl);//Prescribed
flux
    YsBC.valueFraction() = 0.0;//valueFraction must be set to 0
   }
```

[Listing of TStateBC.H]

```
// Computes stoichiometric coefficient, lambda
// Written Steven Beale
// Forschungszentrum Juelich GmbH
// 2017-2020

    label inletPatchID = mesh.boundaryMesh().findPatchID("inlet");
     if (inletPatchID==-1)
    {
      Info << "Failed to find patch inlet"
           <<endl;
    }
    label outletPatchID = mesh.boundaryMesh().findPatchID("outlet");
     if (outletPatchID==-1)
    {
      Info << "Failed to find patch named outlet"
           <<endl;
    }
    // total air mass rate at inlet
    scalarField  rateInlet =
    (
        1.0
        *(
            U.boundaryField()[inletPatchID]
            &
            mesh.Sf().boundaryField()[inletPatchID]
         )
    );

    // total air mass rate at outlet
    scalarField  rateOutlet =
    (
        1.0
        *(
            U.boundaryField()[outletPatchID]
            &
            mesh.Sf().boundaryField()[outletPatchID]
         )
    );

    // total air reaction rate at the membrane
    scalarField  rateCathode =
    (
        kReact.value()
        *(
            pow(y.boundaryField()[membranePatchID],gamma.value())
            *mesh.magSf().boundaryField()[membranePatchID]
         )
```

```
    );

    Info<< "   Total mass rates: [kg/s]" << nl
        << "   inlet: " << Foam::mag(Foam::gSum(rateInlet))
        << "   outlet: " << Foam::mag(Foam::gSum(rateOutlet))
        << "   membrane: " << Foam::mag(Foam::gSum(rateCathode))
        << nl;

    // ----------------------------------------------------------------------

            scalar scalarRateInlet = gSum
            (
                rateInlet*y.boundaryField()[inletPatchID]
            );

            scalar scalarRateOutlet = gSum
            (
                rateOutlet*y.boundaryField()[outletPatchID]
            );

            scalar scalarRateElec = gSum
            (
             rateCathode*(tState.value()-y.boundaryField()[membranePatchID])
            );

            Info<<  " Mass rates transferred substance only: [kg/s]: " << nl;
            Info<< "   inlet = " << Foam::mag(scalarRateInlet)
                << "   outlet = " << Foam::mag(scalarRateOutlet)
                << "   chem = " << Foam::mag(scalarRateElec)
                << "   sum = " <<
Foam::mag(scalarRateElec)+Foam::mag(scalarRateOutlet)-
Foam::mag(scalarRateInlet)
                << nl;

            Info<< "       Air stoichiometric factor by chemRate = "
                << Foam::mag(scalarRateInlet/scalarRateElec)
                << nl
                << "       Air stoichiometric factor by outRate  = "
                << Foam::mag(scalarRateInlet)/(Foam::mag(scalarRateInlet)-
Foam::mag(scalarRateOutlet))
                << nl;
            scalar mlambda =
ilambda.value()*Foam::mag(scalarRateInlet/scalarRateElec);// Use the
electrical stoichiometric factor
            Info<< "   ratio = " << (lambda.value()/mlambda)
                << nl;
            Info<< endl;
// Adjust the velocity if lambda constant
      if (constlambda){
    scalar U1 = Foam::mag
    (
        Foam::gAverage
        (
            U.boundaryField()[inletPatchID]
        )
    );
    Info<< "U1   = "<<U1<<endl;
// Adjust inlet boundary values if using piso
      if (lpiso)
{
```

## lambda.H

```
        U.boundaryFieldRef()[inletPatchID] ==
(1+relax.value()*(lambda.value()/mlambda-
1))*(U.boundaryFieldRef()[inletPatchID]);
            Info<< "   scale factor = " <<
(1+relax.value()*(lambda.value()/mlambda-1))
                << nl;
            Info<< endl;
}
// Adjust internal field values if not using piso
      else {
            U.ref() == U()*(1+relax.value()*(lambda.value()/mlambda-1));
            U.correctBoundaryConditions();
      }
}
```

[Listing of `lambda.H`]

```
// Computes current density, j
// Written Steven Beale
// Forschungszentrum Juelich GmbH
// 2017-2020

//       const dimensionedScalar F
        const scalar F
     (
//          "F",
//          dimensionSet(0, 0, 1, 0, -1, 1, 0),
         96485.3399*1e3    //[C/kmol]
     );
        const scalar rho
     (
//          "rho",
//          dimensionSet(1, -3, 0, 0, 0, 0, 0),
         1.0    //[kg/m^3]
     );
        const scalar MH2
     (
//          "MH2",
//          dimensionSet(1, 0, 0, 0, -1, 0, 0),
         2.0    //[kg/kmol]
     );
        const scalar MO2
     (
//          "MO2",
//          dimensionSet(1, 0, 0, 0, -1, 0, 0),
         32.0    //[kg/kmol]
     );
      if (membranePatchID==-1)
     {
```

## j.H

```
      Info << "Failed to find patch named membrane for T-state boundary
condition"
            <<endl;

      }
      else
      {
      scalar yin = Foam::mag
      (
          Foam::gAverage
          (
              y.boundaryField()[inletPatchID]
          )
      );
      volScalarField& Ys = y;
      volScalarField& id = j;
      scalar jmax = ilambda.value()*(4*F*rho*Uin*yin/MO2)*
      (Foam::gSum(mesh.magSf().boundaryField()[inletPatchID]))/
      (Foam::gSum(mesh.magSf().boundaryField()[membranePatchID]));
      Info<< "jmax   = "<<jmax<<endl;
      id = 2*kReact.value()*pow(Ys,gamma.value())*rho*F/MH2;
      Info << "min,mean,max(current density" << "): "
            << Foam::gMin(id) <<" , "
            << Foam::gAverage(id) <<" , "
            << Foam::gMax(id) << endl;
      }
```

[Listing of `j.H`]

## Re.H

The calculation of the Reynolds number, based on inlet conditions, is not a requirement. Rather, it is conducted to verify that a reasonable value of the inlet velocity, `Uin`, corresponding to appropriate values $k_R$ and $\lambda$ have been chosen.

```
// Computes Reynolds number at inlet
// Written Steven Beale
// Forschungszentrum Juelich GmbH
// 2017-2020

    scalar Uin = Foam::mag
    (
        Foam::gAverage
        (
            U.boundaryField()[inletPatchID]
        )
    );
    scalar Re = dHyd.value()*Uin/nu.value();
    Info<< "nu   = "<<nu.value()<<endl;
    Info<< "U    = "<<Uin<<endl;
    Info<< "Re   = "<<Re<<endl;
```

[Listing of `re.H`]

## References

Beale SB, Choi HW, Pharoah JG, Roth HK, Jasak H, Jeon DH (2016) Comput Phys Commun 200:15–26
Beale SB (2004) J Power Sources 128(2):185–192
Beale SB (2005) Int J Heat Mass Tran 48(15):3256–3260
Beale SB (2007) J Fuel Cell Sci Technol 4(1):1–10
Beale SB (2015) Int J Hydrogen Energy 40(35):11641–11650
Beale SB, Pharoah JG, Kumar A (2013) J Heat Transf 135(1):011004
Beale SB, Reimer U, Froning D, Jasak H, Andersson M, Pharoah JG, Lehnert W (2018) J Electrochem En Con Stor 15(4):041008
Beale SB, Zhang S, Andersson M, Nishida RT, Pharoah JG, Lehnert W (2020) Heat and mass transfer in fuel cells and stacks. In: Brian Spalding D, Runchal AK (eds) 50 years of CFD in engineering sciences a commemorative volume in honour. Springer
Berman AS (1953) J Appl Phys 24(9):1232–1235
Blottner FG (1990) J Spacecr Rocket 27(2):113–122
Bockris JOM, Reddy AKN, Gamboa-Aldeco M (2000) Modern electrochemistry, vol 2. Plenum, New York
Boehm BW (1981) Software engineering economics. Prentice-Hall Inc., Englewood Cliffs, p 122
Cao Q (2017) PhD thesis, Mechanical Engineering. RWTH Aachen University, Aachen
Gileadi E (1993) Electrode kinetics for chemists, chemical engineers and materials scientists. Wiley
Greenshields CJ (2018) The OpenFOAM Foundation: https://openfoam.org.
Kays WM, Crawford ME, Weigand B (2005) Convective heat and mass transfer, 4th edn. McGraw-Hill, New York
Kulikovsky AA (2010) Analytical modelling of fuel cells. Elsevier
Kulikovsky AA (2004) Electrochim Acta 49(4):617–625
Kulikovsky AA, Kucernak A, Kornyshev AA (2005) Electrochim Acta 50(6):1323–1333
Pharoah JG, Djilali N, Vickers GW (2000) J Membrane Sci 176(2):277–289
Reimer U, Froning D, Nelissen G, Raymakers LFJM, Zhang S, Beale SB, Lehnert W (2019) ChemEngineering 3(4):85
Roache PJ, AIAA J (1998) 36(5):696–702
Sherwood TK, Brian PLT, Fisher RE (1963) Desalination Research Laboratory Report 295–301, Massachusetts Institute of Technology, Cambridge
Sherwood TK, Brian PLT, Fisher RE, Dresner L (1965) Ind Eng Chem Fundam 4(2):113–118
Spalding DB (1960) Int J Heat Mass Tran 1:192–207
Spalding DB (1963) Convective mass transfer; an introduction, vol 448. Edward Arnold, London
Zhang S (2019) PhD thesis, Mechanical Engineering. RWTH Aachen, Aachen
Zhang S, Beale SB, Reimer U, Nishida RT, Andersson M, Pharoah JG, Lehnert W (2018) ECS Trans 86(13):287–300

# Low-Temperature Polymer Electrolyte Fuel Cells

**Shidong Zhang**

## 1 Introduction

As a clean and quiet device, polymer-electrolyte fuel cells (PEFCs), also known as proton exchange membrane fuel cells (PEMFCs), have attracted increasing attention over the past decade. PEFCs have shown potential in automotive, backup power, and portable applications due to their low operating temperature, high power density, and quick dynamic response (Andersson et al. 2016; Zhang and Jiao 2018a). Investigations of PEFCs encompass many aspects, such as mechanical stress/compression analysis (Khetabi et al. 2019; Dafalla and Jiang 2018; Qiu 2019), material science (Karan 2017; Zamel 2016; Park et al. 2012), flow-field design (Wang 2015; Arvay et al. 2013), microscale modeling (Fadzillah et al. 2017; Liu et al. 2015; Lile and Zhou 2015), macroscale modeling (Andersson et al. 2016; Zhang and Jiao 2018a; Weber et al. 2014; Wu 2016), etc. Experimental work is significantly advanced with the development of new technologies, such as magnetic resonance imaging (Teranishi et al. 2006), neutron radiography (Panchenko et al. 2018), S++ current scan shunts (Jabbour et al. 2015), and X-ray computed tomography (Jinuntuya et al. 2018). However, conducting such experiments is relatively expensive; in the meantime, comprehensive measurements are usually too expensive. Therefore, mathematical models provide opportunities for researchers to more easily obtain these data.

However, the length scale characteristic of physical processes vary from the microcosmic (nm) to cell/system scale (cm) (Andersson et al. 2016; Zhang and Jiao 2018a;

The original version of this chapter was revised: The ESM material has been updated. The correction to this chapter is available at https://doi.org/10.1007/978-3-030-92178-1_10

**Supplementary Information** The online version contains supplementary material available at https://doi.org/10.1007/978-3-030-92178-1_8.

S. Zhang (✉)
Forschungszentrum Jülich GmbH, Institute of Energy and Climate Research (IEK), 52425 Jülich, Germany
e-mail: s.zhang@fz-juelich.de

© Springer Nature Switzerland AG 2022, corrected publication 2022

S. Beale and W. Lehnert (eds.), *Electrochemical Cell Calculations with OpenFOAM*, Lecture Notes in Energy 42, https://doi.org/10.1007/978-3-030-92178-1_8

Weber et al. 2014). The gas/liquid two-phase flow, together with the electrochemical reaction, mass and heat transfer, electrons and protons transport, water condensation and evaporation, etc., take place within the flow paths and porous media, including the gas diffusion and catalyst layers. Mathematical models that account for different scales of physical processes have also been developed. The origin of PEFC models date back to the 1990s, when Springer et al. (Springer et al. 1991) and Bernardi and Verbrugge (Bernardi and Verbrugge 1992) proposed one-dimensional (1-D) and steady-state mathematical models. These considered water transport in the membrane and predicted overall cell performance. The later work by Nguyen et al. (Nguyen and White 1993; Yi and Nguyen 1998), Gurau et al. (Gurau et al. 1998), and Djilali et al. (Berning et al. 2002; Berning and Djilali 2003) extended the scope of the numerical models. Computational fluid dynamics (CFD) was increasingly applied in PEFC simulations, with the goal of taking major transport phenomena into account.

Water management is one of the most important issues in PEFC design. Two-phase flow, which is common in PEFCs given their low operating temperatures, should be modeled with the goal of achieving better strategies during water management. The volume of fluid (VOF) method was considered powerful due to its capability of tracking gas/liquid interfaces (Zhang and Jiao 2018a; Weber et al. 2014; Ferreira et al. 2015, 2017; Andersson et al. 2019, 2018). However, a significantly high-resolution grid should suffice for the numerical simulation, which limits its applicability. Additionally, it is often used to account for the two-phase flow in flow channels by neglecting other regions and physical processes. Further simplifications are necessary to decrease computational effort and couple other physical processes. It is typically assumed that liquid water in flow channels is in a mist state (Zhang and Jiao 2018b) and/or the liquid phase velocity is an algebraic function of the gas phase velocity (Fan et al. 2017). However, few studies have applied the Eulerian-Eulerian approach (Zhang and Jiao 2018b; Gurau et al. 2008) to address gas and liquid two-phase flows in flow channels.

The majority of the numerical models of PEFC applications were either implemented on the basis of commercial software or developed in-house (Weber et al. 2014). The source codes of these are usually not publicly accessible. Therefore, new features for the implementation and/or model immigrations are problematic, which limits communications between researchers in the fuel cell community. In addition, many numerical simulations are computationally-expensive (Zhang 2019) and require high-performance-computing (HPC) facilities. Open-source codes/packages offer more flexibility in model development/implementation, as well as numerical simulations.

However, there are still limited numbers of open-source packages available nowadays due to the difficulties of model development. The open-source package, openFuelCell (Beale et al. 2016), was developed by Beale et al. (Beale et al. 2016) using the platform of an open-source library, namely OpenFOAM®, with a focus on the CFD modeling of solid-oxide fuel cells (SOFCs). It considers the major transport processes, including heat and mass transfer, charge transfer, electrochemical reaction, and multi-species transfer, by utilizing a multi-region technique. This model was later extended to conduct simulations of high-temperature PEFCs (Zhang 2019), as well as electrochemical hydrogen purification and compression cells (Reimer, et al. 2019). In order to take two-phase flow into account, this model was further developed by the present author and colleagues.

This chapter introduces a new PEFC model and its implementation using OpenFOAM. This model derives from "openFuelCell," with additional consideration of two-phase flow, water transfer in the membrane electrode assembly (MEA), and proton/electron transfer. An Eulerian-Eulerian method is applied to address gas/liquid two-phase flow interactions in terms of momentum and condensation/evaporation.

In order to exhibit the capability of the present solver, studies are conducted to investigate the effects of catalyst cracks. Figure 1 shows the cracks that formed in an in-house-manufactured catalyst layer. Tsushima et al. (2015) reviewed the issues of crack and interfacial voids in PEFCs. They proposed cracks and voids may deteriorate cell performance in light of the larger contact resistances in MEAs. Meanwhile, they also noted that cracks contribute to higher liquid water contents and better gas transport. However, few continuum models were used to consider the influences of cracks and voids.

**Fig. 1** The cracks in a catalyst layer

# 2 Numerical Procedure

## 2.1 Assumptions

- The simulation is steady state.
- Gases are incompressible and obey the ideal gas law.
- Flows are laminar.
- Water is produced in the liquid state.
- Membranes are only permeable to water and protons.
- Gases dissolved in the liquid water are not considered.
- The gas diffusion layers (GDLs) are homogeneous.

## 2.2 Computational Domain

Similar to the approach adopted in 'openFuelCell,' the solver discussed here applies a multi-region technique to address transfer phenomena in different regions, e.g., air, fuel, membranes, etc. In a PEFC, it consists of two main components, namely the interconnect and membrane electrode assembly (MEA). Flow paths are meshed into one/two sides of the interconnect. The MEA can be separated into five distinct layers: (1) The anode gas diffusion layer (aGDL); (2) the anode catalyst layer (aCL); (3) the membrane; (4) the cathode catalyst layer (cCL); and (5) the cathode gas diffusion layer (cGDL).

The simplified computational domain is shown in Fig. 2. It displays a cross-section of a PEFC with a straight channel and its sub-regions. These are used to consider different physical transfer processes:

- Solid region: Only heat transfer is taken into account.
- Fluid region: Single/two-phase flow, species transfer, and heat transfer are considered.
- Electric potential region: electrochemical reactions and electrical fields are addressed.

## 2.3 Governing Equations

### 2.3.1 Fluid Region

Fluid Flow

As is shown in Fig. 2b, the fluid region includes flow channels and porous regions. If the PEFC operates under ambient pressure, water exists in vapor and/or liquid phases.

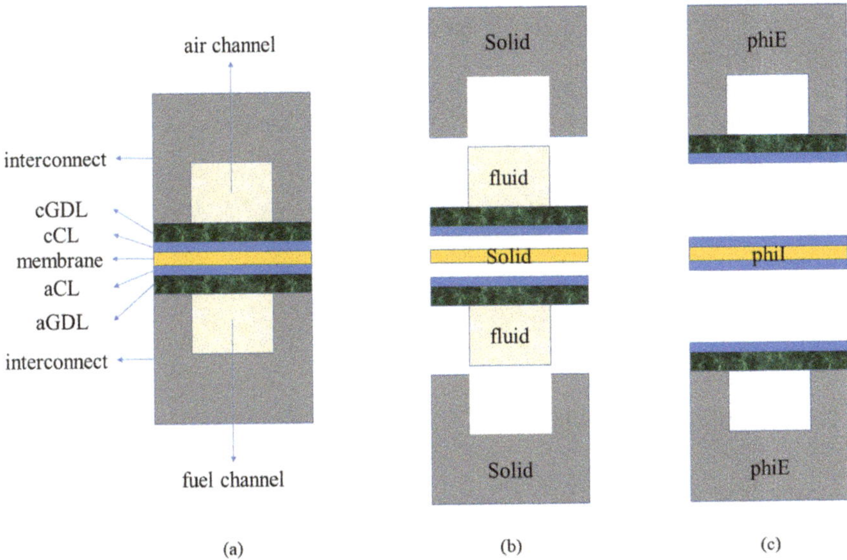


**Fig. 2** Computational domain: **a** Master region; **b** sub-regions: solid and fluid; and **c** sub-regions: electron and proton potentials

Hence, two-phase flow plays an important role in fuel cell performance. The two-phase Eulerian–Eulerian algorithm has been considered by many researchers (Harlow 2004; Harlow and Amsden 1975; Spalding 1981; Ishii and Hibiki 2010; Rusche 2003; Marschall 2011) in the past decades. Based on previous work, the conditionally-averaged continuity and momentum equations can be written as follows:

$$\frac{\partial \rho_\varphi s_\varphi}{\partial t} + \nabla \cdot \left(\rho_\varphi s_\varphi \mathbf{U}_\varphi\right) = R_\varphi \tag{1}$$

$$\frac{\partial \rho_\varphi s_\varphi \mathbf{U}_\varphi}{\partial t} + \rho_\varphi s_\varphi \mathbf{U}_\varphi \cdot \nabla\left(\mathbf{U}_\varphi\right) = -s_\varphi p_\varphi + \nabla \cdot \left(s_\varphi \mu_\varphi \nabla \mathbf{U}_\varphi\right) + \rho_\varphi s_\varphi \mathbf{g} + \mathbf{M}_\varphi + S_{\text{Darcy},\varphi} \tag{2}$$

where $\varphi$ represents the index of phases, $s$ is the phase saturation (volume fraction), $\mathbf{U}$ denotes the phase velocity, $p$ is the phase pressure, $\mu$ is the viscosity, $\mathbf{g}$ represents the gravity, and $\mathbf{M}$ denotes the interface momentum term between each phase. $R$ is the mass source/sink due to condensation/evaporation and electrochemical reactions, and $S_{\text{Darcy}}$ is calculated from Darcy's law as:

$$S_{\text{Darcy}} = \frac{s_\varphi \mu_\varphi}{K_{\text{rel},\varphi} K} \mathbf{U}_\varphi \tag{3}$$

where $K$ is the permeability in the porous materials and $K_{rel}$ stands for the relative permeability that is a function of phase saturation $s$.

For the momentum transfer term, the drag force is usually taken into consideration. For the drag coefficient, a commonly applied model was proposed by Schiller and Naumann (1933):

$$C_d = \frac{24\left(1 + 0.15\mathrm{Re}^{0.687}\right)}{\mathrm{Re}}, \quad \mathrm{Re} \leq 1000 \tag{4}$$

where Re is the Reynolds number.

### Species Transfer

During PEFC operation, hydrogen and air must be humidified in order to obtain sustainable and reliable performance outputs. Multiple components can be observed on both the anode and cathode sides. Meanwhile, water transfers through the Nafion membrane serve as an additional mass source/sink for air and fuel gases. Therefore, the species transfer equations are expressed as:

$$\frac{\partial \rho_\varphi s_\varphi Y_\mathrm{i}}{\partial t} + \nabla \cdot \left(\rho_\varphi s_\varphi \mathbf{U}_\varphi Y_\mathrm{i}\right) = \nabla \cdot \left(\rho_\varphi s_\varphi D_\mathrm{i}^{\mathrm{eff}} \nabla Y_\mathrm{i}\right) + R_\mathrm{i} \tag{5}$$

where $Y$ represents the mass fraction of species i, $R$ is the source/sink term, and $D^{\mathrm{eff}}$ means the effective diffusion coefficient. Readers can refer to the work of Beale et al. (Beale et al. 2016) for the calculation of $D^{\mathrm{eff}}$. It should be noted that gases dissolved in liquid water are not considered in this study. Furthermore, the species transfer equations are only solved in the gas phase. The source/sink term can be expressed as:

$$R_\mathrm{i} = \frac{\omega_\mathrm{i} M_\mathrm{i} J}{z_\mathrm{i}} \tag{6}$$

where $z$ is the number of transferred electrons, $M$ is the molar mass, $\omega$ denotes the stoichiometric coefficient, and $J$ is the volumetric current density or electrochemical reaction rate. The term $J$ is obtained from the 'electric region;' see Sect. 1.3.3.

Considering the two-phase flow in a PEFC, water is subject to condensation and evaporation between the gas and liquid phases.

### Heat Transfer

Heat transfer is an important subject in PEFC simulations. Heat is produced/consumed in many physical processes, such as electrochemical reactions, condensation/evaporation, and electron/proton transfer. In order to address heat transfer in air/fuel streams, an enthalpy equation is commonly used in OpenFOAM applications, shown as:

$$\frac{\partial \rho_\varphi s_\varphi \left(h_\varphi + K\right)}{\partial t} + \nabla \cdot \left(\rho_\varphi s_\varphi \mathbf{U}_\varphi \left(h_\varphi + K\right)\right) = \nabla \cdot \left(\rho_\varphi s_\varphi \alpha^{\text{eff}} \nabla h_\varphi\right) + Q_\varphi \quad (7)$$

where $h$ represents the sensible enthalpy, $\alpha^{\text{eff}}$ is the effective thermal diffusivity, $K$ means mechanical work, and $Q$ encompasses the total heat source/sink. Note that this equation is only solved for one phase; the other phase is considered in the global temperature/enthalpy equation.

### 2.3.2 Solid Region

In the present model, heat transfer is the only process to be considered in solid regions, as shown in Fig. 2b. However, the equations are assembled into the enthalpy equation of the master region. Therefore, no equations are solved in the solid regions. They are only kept for storage of material properties.

### 2.3.3 Electric Region

Potential Fields

The electric region includes two different regions, electronic and ionic (protonic), as shown in Fig. 2c. The governing equations for electronic/ionic potentials have the same formulation:

$$\begin{aligned} \nabla \cdot (\sigma_E \nabla \phi_E) &= J_E \\ \nabla \cdot (\sigma_P \nabla \phi_P) &= J_P \end{aligned} \quad (8)$$

where $\sigma$ is the electric conductivity, $\phi$ is the potential, and $J$ represents volumetric the current density in electrochemical reactions.

The electronic conductivity, $\sigma_E$, in ‘phiE’ (see Fig. 2c), is usually assumed as uniform. However, the protonic conductivity, $\sigma_P$, is a strong function of the water content:

$$\sigma = \begin{cases} (0.514\lambda - 0.326)\exp\left(1268\left(\frac{1}{303} - \frac{1}{T}\right)\right) & 1 \le \lambda \\ 0.1789\lambda \exp\left(1268\left(\frac{1}{303} - \frac{1}{T}\right)\right) & \lambda < 1 \end{cases} \quad (9)$$

where $T$ is the temperature and $\lambda$ is the water content in the membrane.

The electrochemical reactions take place in the catalyst electrodes. The reaction rates, or current density, are described by either Tafel or Butler-Volmer relations:

$$
\begin{aligned}
J_a &= i_{a,0} s_g^{\beta} \prod_i \left(\frac{C_i}{C_{\mathrm{ref},i}}\right)^{\gamma} \left[\exp\left(\frac{\alpha_a z F \eta_a}{R_g T}\right) - \exp\left(-\frac{(1-\alpha_a) z F \eta_a}{R_g T}\right)\right] \\
J_c &= i_{c,0} s_g^{\beta} \prod_i \left(\frac{C_i}{C_{\mathrm{ref},i}}\right)^{\gamma} \left[\exp\left(-\frac{\alpha_c z F \eta_c}{R_g T}\right) - \exp\left(\frac{(1-\alpha_c) z F \eta_c}{R_g T}\right)\right]
\end{aligned}
\tag{10}
$$

where $i_0$ is the exchange current density, $C$ denotes the species concentration in the gas phase, $\gamma$ is the reaction order, $\alpha$ represents the transfer coefficient, $z$ is the number of electron transfers, and $\eta$ means the activation overpotential. The calculations of activation overpotential, $\eta$, refer to:

$$
\begin{aligned}
\eta_a &= \phi_E - \phi_P - E_{n,a} \\
\eta_c &= \phi_E - \phi_P - E_{n,c}
\end{aligned}
\tag{11}
$$

where $E_n$ is the Nernst potential. These formulations are only valid in the catalyst layers of both sides. The Nernst potentials can be written as:

$$
\begin{aligned}
E_{n,a} &= \sum_i \frac{-G_{i,a}\omega_i - R_g T \ln p_{i,a}^{\omega_i}}{zF} = \sum_i \frac{-(H_{i,a} - T S_{i,a})\omega_i - R_g T \ln p_{i,a}^{\omega}}{zF} \\
E_{n,c} &= \sum_i \frac{-G_{i,c}\omega_i - R_g T \ln p_{i,c}^{\omega}}{zF} = \sum_i \frac{-(H_{i,c} - T S_{i,c})\omega_i - R_g T \ln p_{i,c}^{\omega_i}}{zF}
\end{aligned}
\tag{12}
$$

where $G$ is the Gibbs free energy, $H$ is the enthalpy, $S$ is the entropy, and $\omega$ the stoichiometric coefficient.

Then, the relations between the volumetric current density, $J$, and the reaction rates, $J_a$ and $J_c$, can be expressed as the following:

- Cathode catalyst layer:

$$
\begin{aligned}
J_E &= -J_c \\
J_P &= J_c
\end{aligned}
\tag{13}
$$

- Anode catalyst layer:

$$
\begin{aligned}
J_E &= J_a \\
J_P &= -J_a
\end{aligned}
\tag{14}
$$

It can be seen that the above formulae are fairly similar between the cathode and anode sides. Hence, it is straightforward to implement the models into the solver. Finally, the calculation of the surface current density is written as:

$$\mathbf{i} = -\sigma \nabla \phi \tag{15}$$

Dissolved Water

The Nafion membrane must be readily humidified in order to allow proton transfer. The pore sizes of the membrane are of the order of approximately 10 nm. In such a case, water molecules are often localized and not connected in the membrane. The water is assumed to remain in the dissolved phase (Wu et al. 2009). The mechanism of water transfer in the membrane includes diffusion, electro-osmotic drag (EOD), and hydraulic permeation. Of these, the hydraulic permeation can usually be neglected. The non-equilibrium form of water transfer in the membrane yields:

$$\frac{\rho_m}{\mathrm{EW}}\frac{\partial \lambda}{\partial t} + \nabla \cdot \left(\frac{\mathbf{i}}{F} n_d\right) = \frac{\rho_m}{\mathrm{EW}} \nabla \cdot \left(D_m^{\mathrm{eff}} \nabla \lambda\right) + R_\lambda \tag{16}$$

where $\rho_m$ represents the density of the membrane, EW is the equivalent weight, $D_m^{\mathrm{eff}}$ denotes the effective diffusion coefficient in the membrane (Springer et al. 1991):

$$D_m^{\mathrm{eff}} = \begin{cases} 2.693 \times 10^{-10} & \lambda < 2 \\ 10^{-10} \exp\left[2416\left(\frac{1}{303} - \frac{1}{T}\right)\right]\left[0.87(3-\lambda) + 2.95(\lambda - 2)\right] & 2 \le \lambda < 3 \\ 10^{-10} \exp\left[2416\left(\frac{1}{303} - \frac{1}{T}\right)\right]\left[2.95(4-\lambda) + 1.64(\lambda - 3)\right] & 3 \le \lambda < 4 \\ 10^{-10} \exp\left[2416\left(\frac{1}{303} - \frac{1}{T}\right)\right] \times\left[2.563 - 0.33\lambda + 0.0264\lambda^2 - 0.000671\lambda^3\right] & 4 \le \lambda \end{cases} \tag{17}$$

$n_d$ is the drag coefficient that can be expressed as a function of water content (Springer et al. 1991):

$$n_d = \frac{2.5\lambda}{22} \tag{18}$$

and the water absorption/desorption, $R_\lambda$, is obtained from the real and equivalent water contents:

$$R_\lambda = \frac{\rho_m}{\mathrm{EW}} \psi (\lambda - \lambda_e) \tag{19}$$

where $\psi$ is the absorption/desorption coefficient and the equivalent is calculated as (Springer et al. 1991):

$$\lambda_e = \begin{cases} 0.043 + 17.81a - 39.85a^2 + 36.0a^3 & 0 \le a < 1 \\ 14.0 + 1.4(a - 1.0) & 1 \le a < 3 \\ 15.8 & 3 \le a \end{cases} \tag{20}$$

where $a$ represents the water activity:

$$a = \frac{p_{\mathrm{H_2O}}}{p_{\mathrm{sat}}} + 2s \tag{21}$$

### 2.3.4 Master Region

In a PEFC, temperature is the only parameter that must be calculated in all regions. To simplify the case, the temperature field is solved globally in the master region (Beale et al. 2016). The governing equation yields:

$$\frac{\partial \rho C_p T}{\partial t} + \rho C_p \mathbf{U} \cdot \nabla T = \nabla \cdot \left(k^{\mathrm{eff}} \nabla T\right) + Q \tag{22}$$

where $C_p$ represents the specific heat, $Q$ refers to the heat source/sink in each region, $k$ denotes the thermal conductivity, and $\mathbf{U}$ is the moving velocity that is always zero in solid parts. The calculation of effective thermal conductivity gives:

$$k^{\mathrm{eff}} = \varepsilon s_g k_g + \varepsilon s_l k_l + (1 - \varepsilon) k_{\mathrm{porous}} \tag{23}$$

where $\varepsilon$ is the porosity in the porous materials.

## 2.4 Model Implementation

The present model was implemented in the open-source library, OpenFOAM® (https://cfd.direct/openfoam/user-guide). Readers can refer to the home pages of OpenFOAM (www.openfoam.com or www.openfoam.org) for more details. All the governing equations are coupled, discretized, and solved with the finite volume method. The "runTimeSelectionTable" feature, as a built-in function of OpenFOAM, enables highly compact and efficient code structures to be employed. This promotes faster, easier, and more powerful model implementations. The present model enables users to select different solution algorithms for different types of regions, e.g., air side as two-phase flow, and the fuel side is a single-phase flow.

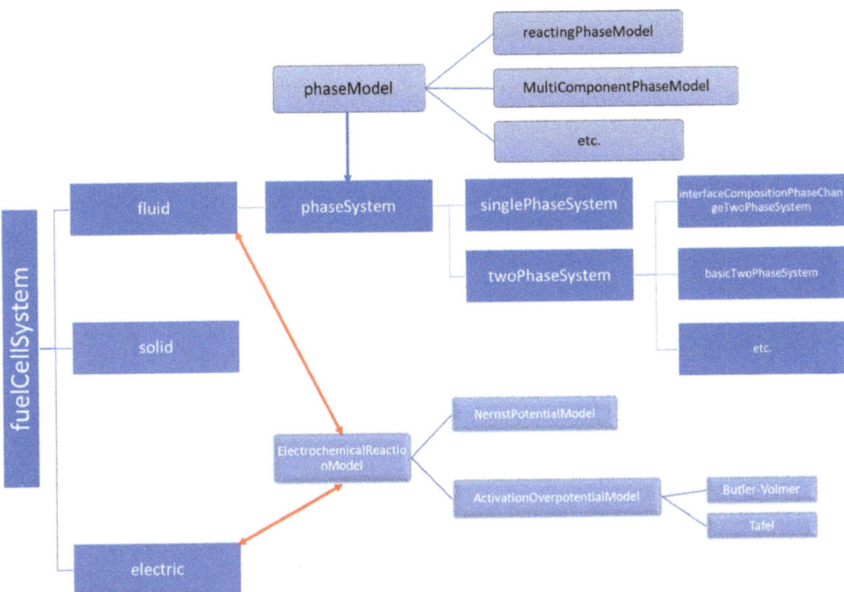


**Fig. 3** Main class structure of the model implementation

The overall class structure of the present model implementation can be seen in Fig. 3. The 'runTimeSelectionTable' is widely used for more flexible applications. The 'parent' and 'child' mesh system, which was first developed by Beale et al. (2016), constructs the top level structure in this implementation. The top-level class, fuelCellSystem, refers to the 'parent' mesh and defines global parameters related to temperature, and local profiles of the 'child' meshes. In addition, it defines the map functions from the 'child' meshes to the 'parent' mesh, and vice versa. The model implementation includes three types of 'child' meshes, namely 'fluid', 'solid', and 'electric', corresponding to the region type, see Fig. 2, and governing equations introduced in the previous section.

**Code example (constant/regionProperties)**

```
// This is the file to define the sub-meshes (also refers to 'child' meshes)
// The regions are defined with a HashTable. The key represents the type of sub-meshes, and
the value is a (word) list of names.

regions
(
    fluid (air fuel)
    solid (electrolyte interconnect)
    electric (phiEA phiEC phiI)
);

Note: this style is similar to the built-in solver 'chtMultiRegionFoam'. In the present
code, the regions are defined in a more effective way.
```

The module, ElectrochemicalReactionModel, is specifically created for the applications with electrochemical reactions. It communicates with both the 'fluid' regions and 'electric' regions, which is only valid in overlapped zones, e.g. the catalyst layers in Fig. 2. In this module, the Nernst potential and activation overpotential are carefully taken into consideration. As a subsystem of the fuelCellSystem, it provides the core relations of heat, mass, and charge transfers between the coupled regions. From the coding perspective, the module is implemented as a type of combustionModel that is well designed in OpenFOAM. There is an option to turn on/off the electrochemical reactions, or switch to other combustion models (beyond the scope of this work).

The phaseSystem module is another important part since it accounts for the fluid flow in the system. The submodule, twoPhaseSystem, is derived from the standard solver of the foundation version OpenFOAM, reactingTwoPhaseEulerFoam. With the 'template' function of C ++, various two-phase flow systems have been included, e.g. basicTwoPhaseSystem, and interfaceCompositionPhaseChangeTwoPhaseSystem, etc. The former refers to a two-phase solver that only considers momentum interactions between two phases. The latter accounts additionally for the interfacial heat and mass transfer because of phase change (condensation/evaporation) through a virtual interface between them. The singlePhaseSystem offers the solver for single phase flows. Compared to the twoPhaseSystems, its implementation is much simpler providing the interactions with additional phases are not present. The example of usage can be found in the following.

This section offers a brief overview of the model implementation. To ensure the solver functions properly, additional models, e.g. diffusivity, porosity, conductivity, and drag, etc., have to be included, and many adjustments have to be conducted. Further details of model implementation and tutorials setup will be published in a future paper. Meanwhile, the source code will be open to the public.

## 2.5 Model Parameters

The governing equations shown in the last section represent the major physical processes in a PEFC, including heat and mass transfer, electron and proton transfer, electrochemical reactions, and two-phase flow. Boundary conditions are necessary to determine the numerical solutions in the computational domains. In a PEFC, the boundary conditions refer to the operating conditions, e.g., galvnostatic/potentostatic mode, pressure, temperature, etc. Material properties and geometric information are also important for numerical simulations. Table 1 lists the parameters necessary to conduct the present simulations.

The velocity is calculated from the prescribed stoichiometric factor, $\lambda$,

$$u = \frac{M_{\mathrm{i}} A_{\mathrm{act}} \bar{i} \lambda}{\rho X_{\mathrm{i}} A_{\mathrm{in}} z F} \tag{24}$$

**Table 1** Model parameters and operating conditions

| Description | Value | Unit |
|---|---|---|
| Channel height | 1.0 | mm |
| Channel width | 1.0 | mm |
| Channel length | 24.0 | mm |
| Rib width | 0.5 | mm |
| Rib height | 3.0 | mm |
| Thickness<br>GDL, MPL, CL, membrane | 0.28, 0.02, 0.0129, 0.02 | mm |
| Porosity<br>GDL, MPL, CL | 0.7, 0.4, 0.3 | – |
| Permeability<br>GDL, MPL, CL | 10, 1.0, 0.1 | $\mu m^2$ |
| Contact angle<br>GDL, MPL, CL | 120, 120, 100 | – |
| Surface tension<br>GDL, MPL, CL | 0.625, 0.625, 0.625 | $N \cdot m^{-1}$ |
| Thermal conductivity<br>BPP, GDL, MPL, CL, membrane | 20, 1.0, 1.0, 1.0, 0.95 | $W\,m^{-1}K^{-1}$ |
| Electric conductivity<br>BPP, GDL, MPL, CL | 20,000, 8000, 8000, 5000 | $S\,m^{-1}$ |
| Water absorption/desorption rate | 1.3, 1.3 | $s^{-1}$ |
| Droplet diameter | 0.15 | mm |
| Exchange current density (a/c) | $1.0 \times 10^8/120$ | $A\,m^{-3}$ |
| Transfer coefficient (a/c) | 0.5/0.5 | – |
| Pressure (a/c) | 101,325/101325 | Pa |
| Temperature (a/c) | 353.15/353.15 | K |
| Stoichiometric factor (a/c) | 2/2 | – |
| Relative humidity (a/c) | 50%/50% | – |

* Note: a: anode; c: cathode

**Table 2** Model parameters in the GDLs for validation

| Description | Value | Unit |
|---|---|---|
| Current density | 1.4 | $A\,cm^{-2}$ |
| Contact angles | 80<br>100 | o |
| Permeability | $6.875 \times 10^{-13}$ | $m^2$ |
| Porosity | 0.5 | – |
| Thickness | 0.3 | mm |

**Table 3** Operating conditions for validation

| Description | Value | Unit |
|---|---|---|
| Pressure | a: 101,325<br>c: 101,325 | Pa |
| Temperature | 343.15 | K |
| Stoichiometric factor | a: 2<br>c: 2 | – |
| Relative humidity | a: 90%<br>c: 90% | – |

Note: a: anode; c: cathode

where $M$ is the molar mass, $A_{\mathrm{act}}$ denotes the active area, $A_{\mathrm{in}}$ represents the area of the inlet cross-section, $X$ is the molar fraction, and $\bar{i}$ is the mean current density.

A constant temperature applies at the bottom and top surfaces of the PEFC. Meanwhile, the fuel cell operates in galvanostatic mode, i.e., the mean current density is prescribed (as opposed to potentiostatic mode where the potential is fixed).

## 2.6 Computational Domain

The computational geometry considered in this chapter is displayed in Fig. 4. Different components are marked with different colors, which correspond to the

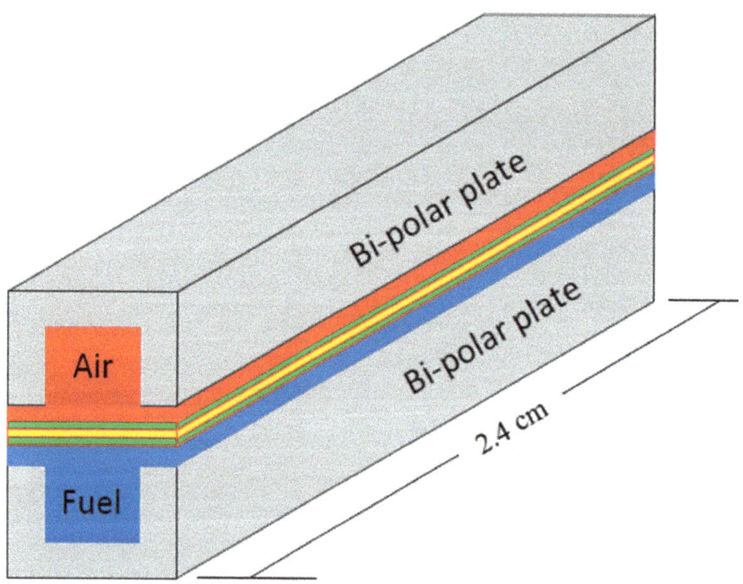


**Fig. 4** Computational geometry (not to scale)

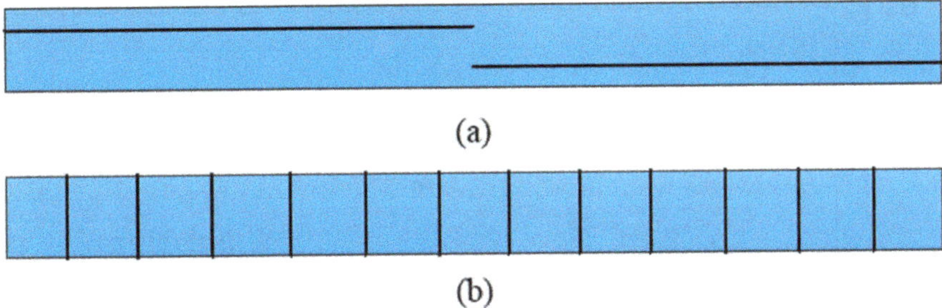


**Fig. 5** Positions and orientations of cracks in the CCLs: **a** Horizontal; **b** vertical (not to scale)

cross-section shown in Fig. 2. The dimensions of each part can be found in Table 1. Figure 5 depicts the locations and orientations of cracks in the CCLs. These are assumed as void spaces, where no electrochemical reactions occur. As suggested by Berg and Kulikovsky (2015) the width of cracks in PEFCs varies from 10–30 $\mu$m. The width of these may increase over long-time operation.

## 2.7 Model Verification and Validation

Before applying the present solver to PEFC simulations, the author performed a verification from the work of Pasaogullari and Wang 2004) and Beale et al. (2009). Using a 1-D model, the liquid water movement from the catalyst layer to the gas channels in a PEFC was studied.

Table 2 shows the parameters for the two cases, namely hydrophobic and hydrophilic GDLs. The comparison between the analytical solutions and simulation results can be seen in Fig. 6. The hydrophobic GDL tends to retain a lower water content. It can also be seen that the deviations between analytical solutions and simulations are fairly minor. Therefore, the present model is able to predict water evolution in porous GDLs.

The validation was conducted using an in-house-designed PEFC prototype. Readers can refer to the work of Shi et al. (2019) for details of the geometry. The polarization curve was measured in galvanostatic mode with the operating conditions shown in Table 3. The comparison of.

polarization curves between the simulation results and experimental measurement is presented in Fig. 7. It can be seen that the deviations between them are minor, with the exception of the open-circuit-voltage condition. In that case, the present model predicts higher cell voltages compared to the experimental data. A detail discussion of the subject was presented in the work of Reimer et al. (2018). In a PEFC, the reactants dissolve in the liquid film for the electrochemical reactions, whereas the present model only considers the gas phase concentrations. Nevertheless, PEFCs typically operate with reasonable currents, in which case the model predicts voltages close to the experiments.

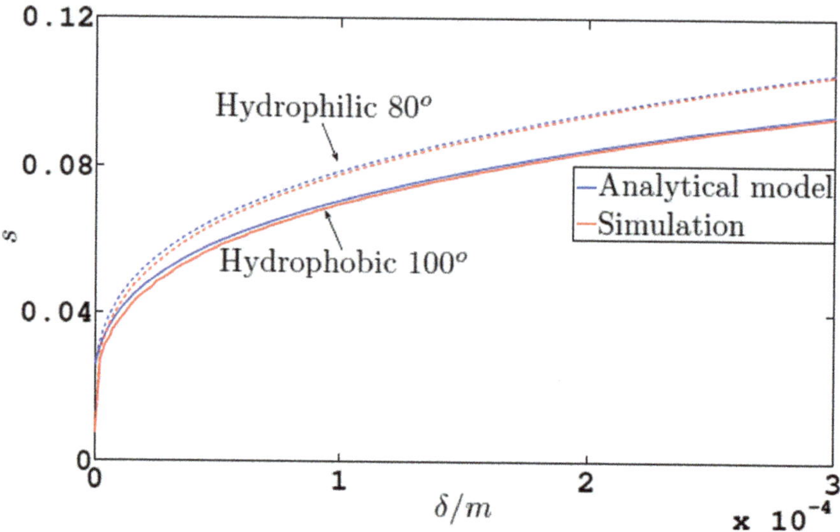


**Fig. 6** Comparison of liquid water saturations in the GDLs (left: channel; right: CL)

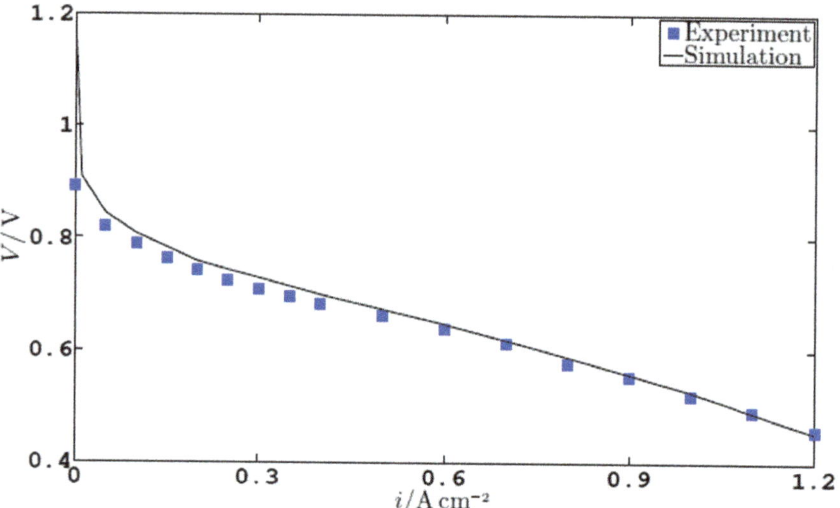


**Fig. 7** Comparison of polarization curves between the simulation and experimental data

## 3 Results

The cracks are only located at the cathodic catalyst electrode and the initial crack width is assumed to be 20 $\mu$m. The orientations of the cracks are depicted in Fig. 5. The number of cracks and their widths may increase after long-term operation.

## 3.1 Initial Crack Width

In order to study the effects arising from catalyst cracks, a case with no cracks was also taken into account. A comparison of polarization curves between cases with/without cracks can be seen in Fig. 8. It is clear that the overall cell performances, or cell voltages, were not affected by the cracks. Considering the electrochemical reactions, the volumes of the cracks occupy 1% of the total reaction zone. However, liquid water should accumulate in these cracks due to the electrochemical reaction. Therefore, the local fields may be altered in the light of the existence of the cracks.

Figure 9 shows the through-plane current density distributions in the middle-plane of the membranes. It can be observed that the global current density distributions

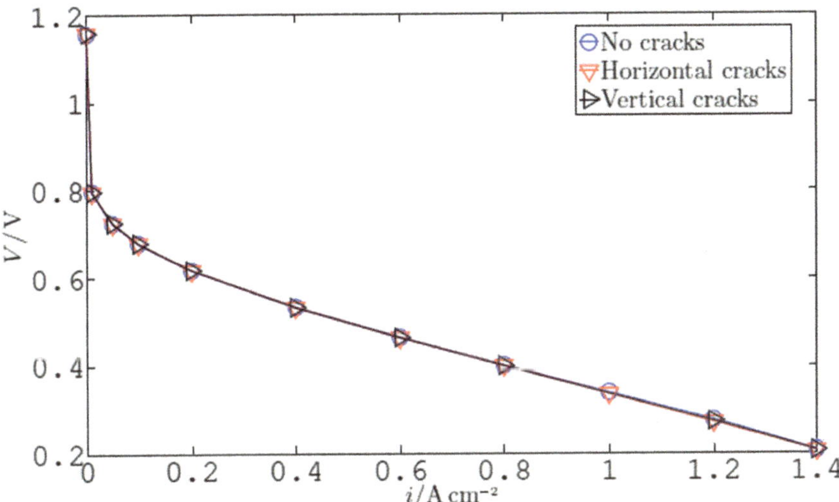


**Fig. 8** Comparison of polarization curves for cases with/without cracks

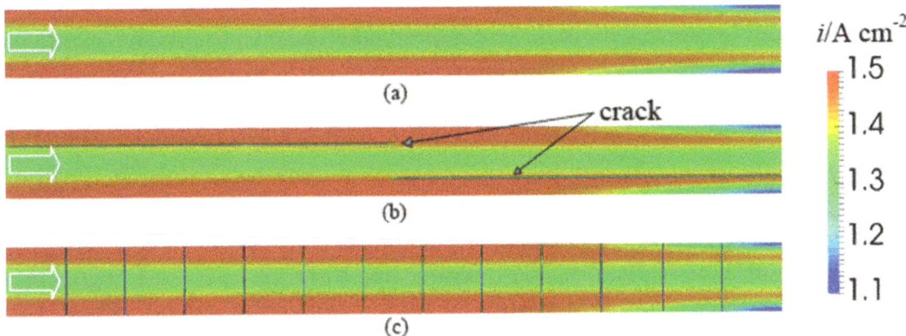


**Fig. 9** Comparison of current density distributions, $i = 1.4$ A cm$^{-2}$. **a** Case without cracks; **b**. case with horizontal cracks; and (c) case with vertical cracks

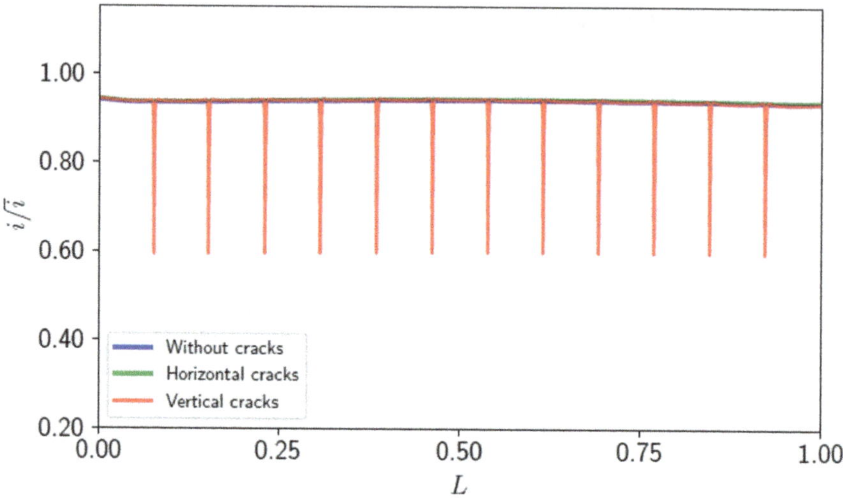


**Fig. 10** Local current density variations, $i = 1.4$ A cm$^{-2}$. (Central line of the membrane from the inlet to the outlet)

were fairly similar. The magnitude of the current density decreases from the inlet to the outlet, and from the land regions to the channel ones. This mainly results from the consumption of reactants and higher water content in the land regions, respectively.

However, the current density is insignificantly small in the crack regions. Figure 10 showed the local current density values on the central lines from the inlet to the outlet. The obvious differences were found in the crack regions, but negligible deviations obtained in the others. The strongly decreased current density mainly resulted from the relatively thin membrane. Proton transfer in the membrane, which corresponds to the current, was virtually two-dimensional. In the crack regions, where electrochemical reactions disappeared, a small number of protons moved across.

Another important parameter in PEFCs is liquid water saturation. Hydrophobic porous electrodes removed additional amounts of liquid water from the cathodic catalysts in order to prevent flooding. As void spaces, the cracks can retain some amount of liquid water. This should be considered and taken into account, especially when the gases have been fully humidified or fuel cells operated with high mean current densities.

The simulation results for cases with/without cracks can be observed in Fig. 11. In this plot, the global distributions of liquid water saturations were nearly identical, and the magnitudes increased from the inlet to the outlet. The electrochemical reaction generated liquid water, which gradually accumulated near the outlet. Figure 12 shows the local liquid water saturations on a central line from the inlet to the outlet. The overall liquid water saturations slightly increased, whereas the local values in the crack regions were significantly higher, at more than 40%. It should be noted that the effects only applied in the crack regions and decreased rapidly to nearby regions. The cracks contributed to higher water contents in the catalyst electrodes.

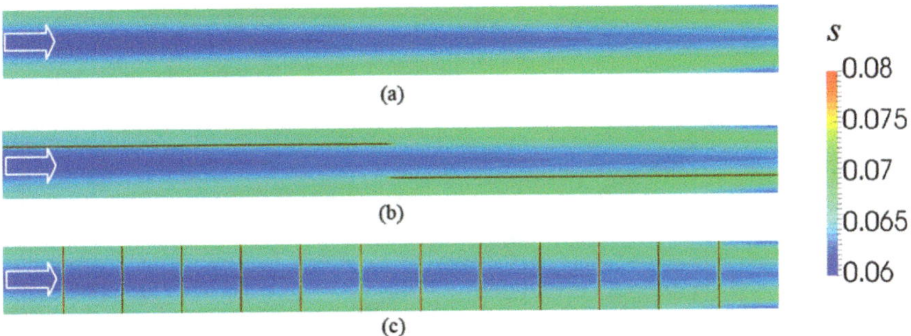


**Fig. 11** Liquid water saturation on the interface between the electrode and membrane, $i = 1.4$ A cm$^{-2}$. **a** Case without cracks; **b** case with horizontal cracks; and **c** case with vertical cracks

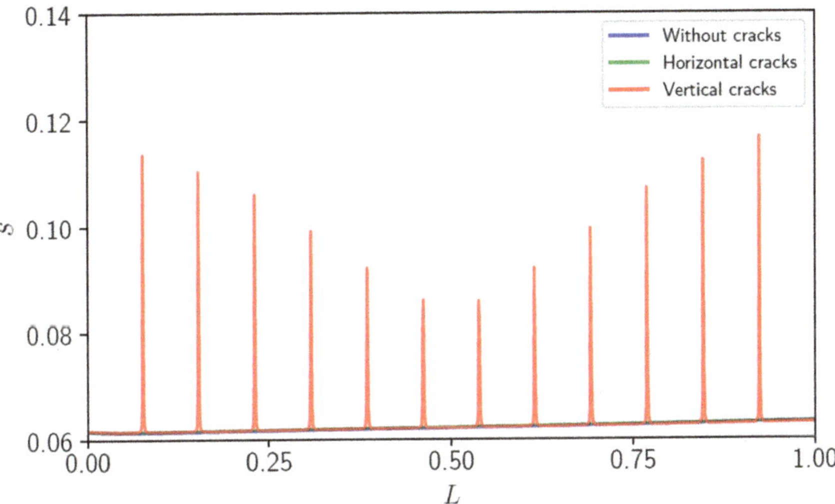


**Fig. 12** Local liquid water saturations, $i = 1.4$ A cm$^{-2}$. (A central line from the inlet to the outlet)

The oxygen mole fraction variations on the central line of the membrane surface are depicted in Fig. 13. The values decreased from the inlet to the outlet due to the electrochemical consumption. It can be seen that cases with cracks presented slightly higher oxygen concentrations than those without them. The maximum difference, of ~ 8%, appeared near the outlet, where the oxygen molar fractions were relatively low. In the regions near to the cracks, a maximum increase of 5% could be observed. Therefore, the cracks contributed to better oxygen diffusion from the porous GDLs to the catalysts.

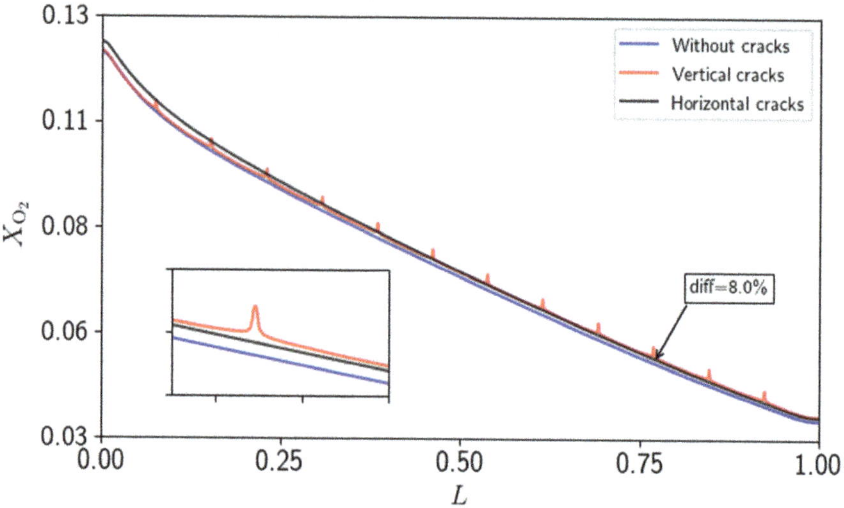


**Fig. 13** Oxygen mole fraction variations from the inlet to the outlet, $i = 1.4\ \mathrm{A\ cm^{-2}}$. (Central line of the interface between the membrane and cathode catalyst electrode)

## 3.2 *Wider Cracks*

The initial width of cracks in previous sections was assumed to be 20 μm. During long-term operations, the width may increase. Therefore, different widths (i.e., 40, 60, or 80 μm) of cracks are analyzed in this section. Simply put, the widths along the cracks were constant.

Figure 14 shows the cell output voltages for cases with different crack widths. It can be seen that the voltages decreased with wider cracks. This resulted from the increase in the missing reaction volumes within the catalyst layers. In the scope of present study, the maximum variation was within 3.5%. Meanwhile, the cases with horizontal cracks provided slightly lower voltages compared to cases with vertical ones.

The current density distributions for cases with 0.06 mm-wide cracks are shown in Fig. 15. The deviations in local current densities between the cracks and nearby regions became more significant. Vertical cracks slightly distorted the distributions of the current density.

The local variations are plotted in Fig. 16, where it can be seen that the minimum current densities, which are located in the crack regions, accounted for approximately 25% of the current densities in nearby regions. Considering the thickness of the membrane, the current flowed two-dimensionally, which meant few protons transferred in the crack regions from the anode to the cathode sides. The magnitude of the current density in this area became lower for larger cracks (not shown here). In addition, slightly higher current densities can also be observed in the adjacent area between the cracks and nearby regions. As the cracks were assumed to be located on

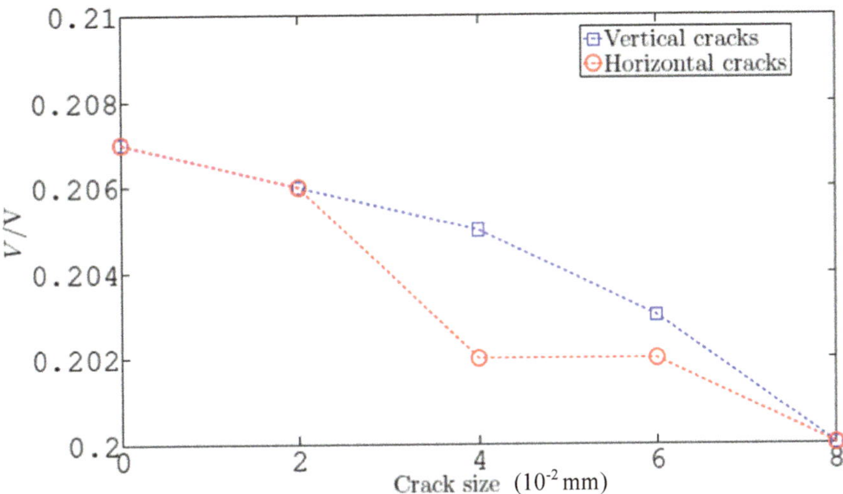


**Fig. 14** Cell voltages for cases with different widths, $i = 1.4$ A cm$^{-2}$

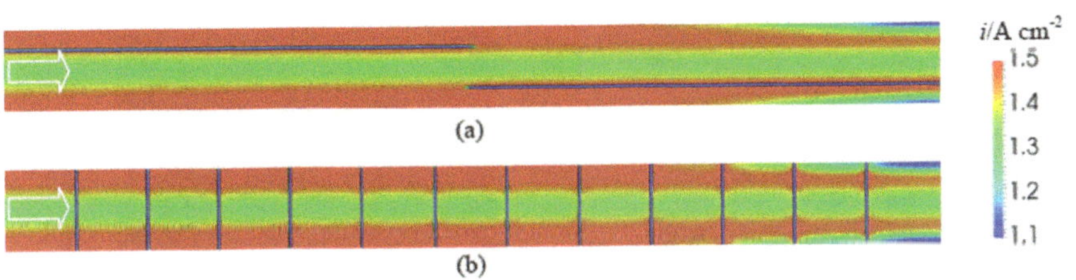


**Fig. 15** Comparison of current density distributions (crack width = 60 μm), $i = 1.4$ A cm$^{-2}$. **a** Case with horizontal cracks; and **b** case with vertical ones

the cathode side, the protons generated on the anode side tended to accumulate near the boundary of the crack regions.

Figures 17 and 18 present the liquid water saturation distributions on the interfaces between the membrane and cathodic catalyst layer and their local variations in the central lines, respectively. In this case, liquid water saturation levels were much higher in the cracks than nearby areas. This indicates that it was easier for liquid water to accumulated in the cracks. Meanwhile, the overall water content in the catalyst layer increased. In Fig. 18, the maximum saturation appeared in the cracks, and was nearly twice the magnitude in nearby areas. In the case of higher relative humidity from inlet gases, the flooding issue would more preferably occur in the cracks.

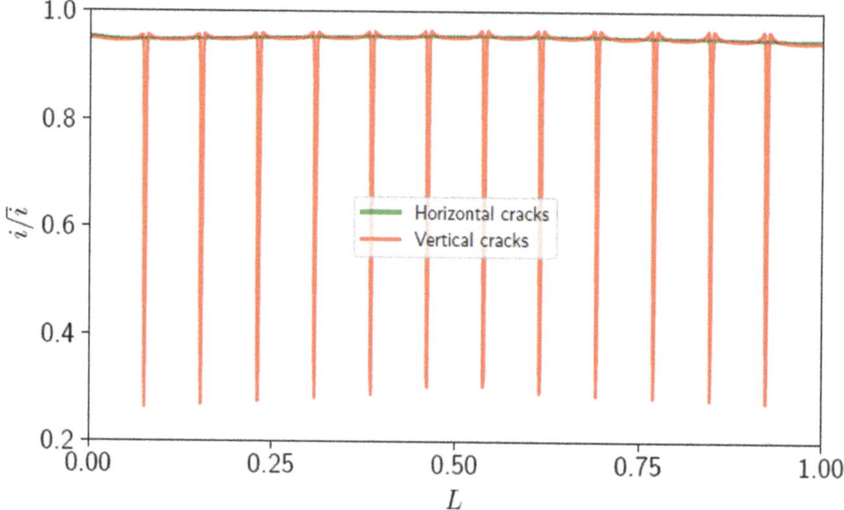


**Fig. 16** Local current density variations (crack width = 60 μm), $i = 1.4$ A cm$^{-2}$

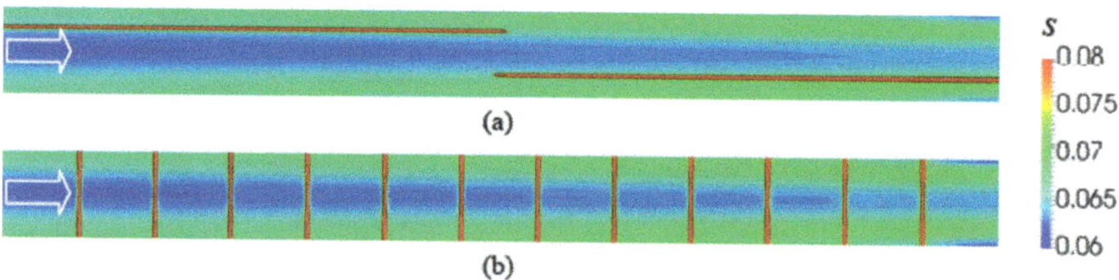


**Fig. 17** Liquid water saturation distributions (crack width = 60 μm), $i = 1.4$ A cm$^{-2}$. **a** Case with horizontal cracks; and **b** case with vertical ones

## *3.3 Discussion*

This chapter introduces a PEFC model, which has been implemented in OpenFOAM. The development of this solver was based on two released repositories, namely openFuelCell (OpenFuelCell) and reacting EulerFoam (the built-in solver in OpenFOAM). The former mainly offered the multiple-region method, whereas the latter was used to solve the two-phase flows. A two-phase Eulerian–Eulerian algorithm is employed to address the interactions between the liquid and gas phases. In the present solver, the liquid water is treated as spherical droplets. The attachment/detachment of droplets from the surfaces of the GDLs and walls are not taken into account. These simplifications apply if the liquid water exists in a mist state within the flow paths; in other words, the liquid water saturation should be relatively low. However, the flooding issue during fuel cell operation is an unresolved problem. In this case, annular/slug flow can occur in the channels and may percolate into the GDLs, which

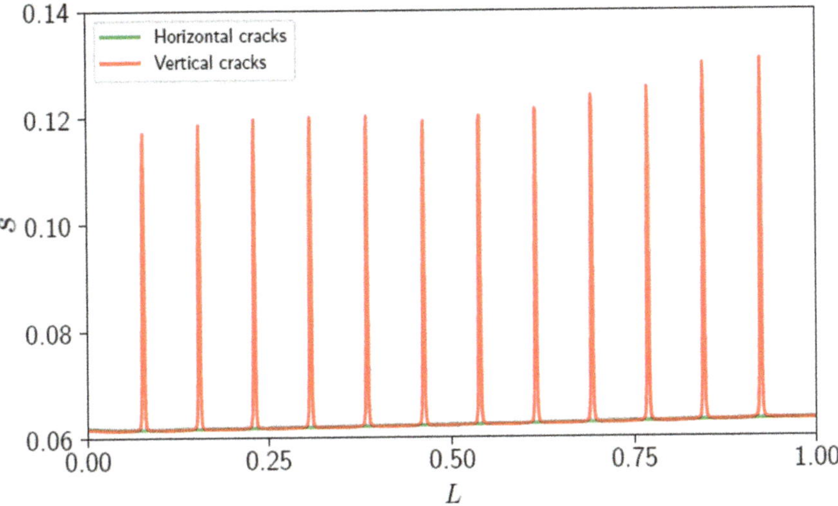


**Fig. 18** Liquid water saturation variations (crack width = 60 μm), $i = 1.4\ \mathrm{A\ cm^{-2}}$

prevents reactants from diffusing to the reaction sites. In such a case, a volume of fluid (VOF) method provides higher-order predictions for consideration of the flooding issue. However, the computational requirement of the VOF method limits the applications at the cell-level. A balance between accuracy and computational requirements needs to be reached. The volume-averaged two-phase Eulerian–Eulerian method in the present model is, in many ways, a compromise for simulating the two-phase flows and coupling the additional physical processes in a PEFC.

Carbon paper/cloth is typically utilized in the porous electrodes (GDL) in a PEFC. Some researchers (Froning et al. 2016; Yu et al. 2018a, b) have applied a Lattice Boltzmann Method (LBM) to investigate two-phase flow in a reconstructed GDL. Detailed CFD simulations of flow through the pores of the porous material are virtually impossible considering the computational requirements, spatial/temporal resolutions, structure reconstructions, mesh generations, etc., though see Beale et al. [ref] Therefore, in cell-level modeling, the GDLs are treated as homogeneous regions, described in terms of permeability, porosity, and tortuosity. In this study, these values are treated as constant over the GDLs. This is not the case in many realistic applications, e.g., some parts of GDLs are compressed during assembly. The mechanical stress contributes to the deformation of GDLs in the rib regions, in which the permeability and porosity decrease, but the tortuosity increases. These values also change after long-term operation. Another issue concerns the contact resistance between the GDLs and solid bipolar plates. The thermal/electrical contact resistances may be expressed as functions of the mechanical stresses (Liang et al. 2018); points that should be addressed in future studies.

In the present model, the membrane does not allow gases to permeate. The driving force for water to move through the membrane includes concentration differences (diffusion) and the drag force due to protons (osmotic drag). However, studies have revealed that the reactants, i.e., hydrogen and oxygen, also diffuse from one side to the other one. The reactant crossover typically leads to accelerated membrane degradation (Nam et al. 2010) and lower cell performance (Chippar and Ju 2013). The situation deteriorates if pinholes (Chippar et al. 2014) are formed in the membranes, as the gas crossover then becomes more significant, especially if high pressure differences exist between the cathode and anode sides. The crossover issue is beyond the scope of this study. However, the cracks formed in the catalyst layer increase the risk of pinhole formation (Kundu et al. 2006). This topic should be further investigated in the future.

In this study, the effects of the cracks were investigated in terms of their geometric aspects. It was found that the influences were fairly small for the overall cell performance and the distributions of several parameters. Although the cracks increased the oxygen molar fractions in the catalyst layers, consistent with the suggestions of Karst et al. (2010), the output voltages were virtually identical. Narrow cracks usually appear in newly-made catalyst layers, which exert no influence on cell performance. However, the cracks contribute to the formation of pinholes. Meanwhile, the membrane and catalyst degradation and delamination take place in the PEFCs after long-term operation (Kim and Mench 2007; Ding et al. 2016; Kim et al. 2009). These combination effects may result in more significant differences in both local and overall performances.

In this study, the multiple region method was applied to account for the coupling processes in different regions. This tends to make the solver slow and unstable during PEFC simulations. In order to increase the stability, several relaxation factors were implemented: higher values contribute to faster convergence, but to lower stability, and vice versa. A balance is needed between stability and simulation speed. In addition, the two-phase flow greatly slows down convergence. Although a local time stepping (LTS) method was utilized for steady-state two-phase flow simulations, the simulation process still remains time-consuming. A superior solution scheme should be strived-for in order to address the convergence issue.

## 4 Conclusions

In this chapter, a 3-D steady-state, non-isothermal, multi-component, two-phase model was introduced for PEFCs. A two-phase Eulerian–Eulerian algorithm was applied to the two-phase flow simulations. The solver enables consideration of the major physical processes in a PEFC, including heat and mass transfer, two-phase flow, electrochemical reactions, and proton/electron transfer. A multiple-region method was used to account for these physical processes in different regions. The model was been implemented using the open-source library, OpenFOAM.

The implemented model was verified by analytical solutions and validated with polarization curves measured from the in-house PEFC prototype. The present model is capable of predicting the liquid water saturations and cell voltages with minor deviations from experimentally-obtained values.

Simulations were conducted to study the effects arising from cracks in the cathodic catalyst layers. Under the operating conditions assumed in this study, the cracks exerted only a slight influence on the overall performances. The local variations of current densities and liquid water saturations were significantly altered. Lower current density and higher water saturation were also observed in the crack regions, which become more apparent for wider cracks. The cracks increased the water content and oxygen molar fraction in the catalyst layers.

**Acknowledgements** The authors appreciate the help from many colleagues in IEK-14, Juelich Research Center. The discussion with Prof. Hrvoje Jasak, Dr. Henrik Rusche, and Dr. Holger Marschall provides many valuable insights for this research. The support of Juelich Aachen Research Alliance (JARA) High Performance Computing (HPC) is important to conduct the numerical simulations. The language review and editing performed by Mr. Christopher Wood is also appreciated.

## References

Andersson M et al (2016) A review of cell-scale multiphase flow modeling, including water management, in polymer electrolyte fuel cells. Appl Energy 180:757–778

Zhang G, Jiao K (2018a) Multi-phase models for water and thermal management of proton exchange membrane fuel cell: A review. J Power Sources 391:120–133

Khetabi EM et al (2019) Effects of mechanical compression on the performance of polymer electrolyte fuel cells and analysis through in-situ characterisation techniques - areview. J Power Sources 424:8–26

Dafalla AM, Jiang F (2018) Stresses and their impacts on proton exchange membrane fuel cells: areview. Int J Hydrogen Energy 43(4):2327–2348

Qiu Det al (2019) Mechanical failure and mitigation strategies for the membrane in a proton exchange membrane fuel cell. Renew Sustain Energy Rev 113: 109289

Karan K (2017) PEFC catalyst layer: recent advances in materials, microstructural characterization, and modeling. Curr Opin Electrochem 5(1):27–35

Zamel N (2016) The catalyst layer and its dimensionality – alook into its ingredients and how to characterize their effects. J Power Sources 309:141–159

Park S, Lee J-W, Popov BN (2012) A review of gas diffusion layer in PEM fuel cells: materials and designs. Int J Hydrogen Energy 37(7):5850–5865

Wang J (2015) Theory and practice of flow field designs for fuel cell scaling-up: a critical review. Appl Energy 157:640–663

Arvay A et al (2013) Nature inspired flow field designs for proton exchange membrane fuel cell. Int J Hydrogen Energy 38(9):3717–3726

Fadzillah DM et al (2017) Review on microstructure modelling of a gas diffusion layer for proton exchange membrane fuel cells. Renew Sustain Energy Rev 77:1001–1009

Liu X et al (2015) Liquid water transport characteristics of porous diffusion media in polymer electrolyte membrane fuel cells: A review. J Power Sources 299:85–96

Lile JRD, Zhou S (2015) Theoretical modeling of the PEMFC catalyst layer: a review of atomistic methods. Electrochim Acta 177:4–20

Weber AZ et al (2014) A critical review of modeling transport phenomena in polymer-electrolyte fuel cells. J Electrochem Soc 161:F1254–F1299

Wu H-W (2016) A review of recent development: transport and performance modeling of PEM fuel cells. Appl Energy 165:81–106

Teranishi K, Tsushima S, Hirai S (2006) Analysis of water transport in PEFCs by magnetic resonance imaging measurement. J Electrochem Soc 153(4):A664–A668

Panchenko O et al (2018) In-situ two-phase flow investigation of different porous transport layer for a polymer electrolyte membrane (PEM) electrolyzer with neutron spectroscopy. J Power Sources 390:108–115

Jabbour L et al (2015) Feasibility of in-plane GDL structuration: Impact on current density distribution in large-area proton exchange membrane fuel cells. J Power Sources 299:380–390

Jinuntuya F et al (2018) The effects of gas diffusion layers structure on water transportation using X-ray computed tomography based Lattice Boltzmann method. J Power Sources 378:53–65

Springer TE, Zawodzinski TA, Gottesfeld S (1991) Polymer electrolyte fuel cell modell. J Electrochem Soc 138(8):2334–2342

Bernardi DM, Verbrugge MW (1992) A mathematical model of the solid-polymer-electrolyte fuel cell. J Electrochem Soc 139(9):2477–2491

Nguyen TV, White RE (1993) A water and heat management model for proton-exchange-membrane fuel cells. J Electrochem Soc 140(8):2178–2186

Yi JS, Nguyen TV (1998) An along–the–channel model for proton exchange membrane fuel cells. J Electrochem Soc 145(4):1149–1159

Gurau V, Liu H, Kakaç S (1998) Two-dimensional model for proton exchange membrane fuel cells. AIChE J 44(11):2410–2422

Berning T, Lu DM, Djilali N (2002) Three-dimensional computational analysis of transport phenomena in a PEM fuel cell. J Power Sources 106:284–294

Berning T, Djilali N (2003) A 3D, multiphase, multicomponent model of the cathode and anode of a PEM fuel cell. J Electrochem Soc 150:A1589–A1598

Ferreira RB et al (2015) Numerical simulations of two-phase flow in proton exchange membrane fuel cells using the volume of fluid method – a review. J Power Sources 277:329–342

Ferreira RB et al (2017) 1D+3D two-phase flow numerical model of a proton exchange membrane fuel cell. Appl Energy 203:474–495

Andersson M et al (2019) Modeling of droplet detachment using dynamic contact angles in polymer electrolyte fuel cell gas channels. Int J Hydrogen Energy 44(21):11088–11096

Andersson M et al (2018) Modeling and synchrotron imaging of droplet detachment in gas channels of polymer electrolyte fuel cells. J Power Sources 404:159–171

Zhang G, Jiao K (2018b) Three-dimensional multi-phase simulation of PEMFC at high current density utilizing Eulerian-Eulerian model and two-fluid model. Energy Convers Manage 176:409–421

Fan L, Zhang G, Jiao K (2017) Characteristics of PEMFC operating at high current density with low external humidification. Energy Convers Manage 150:763–774

Gurau V, Zawodzinski TA Jr, Mann JA Jr (2008) Two-phase transport in PEM fuel cell cathodes. J Fuel Cell Sci Technol 5(2)

Zhang S et al (2019) Polymer electrolyte fuel cell modeling - acomparison of two models with different levels of complexity. J Electrochem Soc

Beale SB et al (2016) Open-source computational model of a solid oxide fuel cell. Comput Phys Commun 200:15–26

Reimer U et al (2019) An engineering toolbox for the evaluation of metallic flow field plates. ChemEngineering 3(4)

Tsushima S, Hiran S (2015) An overview of cracks and interfacial voids in membrane electrode assemblies in polymer electrolyte fuel cells. J Therm Sci Technol 10(1):JTST0002–JTST0002

Harlow FH (2004) Fluid dynamics in group T-3 Los Alamos national laboratory: (LA-UR-03-3852). J Comput Phys 195(2):414–433

Harlow FH, Amsden AA (1975) Numerical calculation of multiphase fluid flow. J Comput Phys 17(1):19–52

Spalding DB (1981) Numerical computation of multi-phase fluid flow and heat transfer. In: Von Karman Institute for fluid dynamics in numerical computation of multi-phase flows, pp 161–191
Ishii M, Hibiki T (2010) Thermo-fluid dynamics of two-phase flow. Springer Science & Business Media
Rusche H (2003) Computational fluid dynamics of dispersed two-phase flows at high phase fractions. Imperial College London (University of London)
Marschall H (2011) Towards the numerical simulation of multi-scale two-phase flows. Technische Universität München
Schiller L (1933) A drag coefficient correlation. Zeit Ver Deutsch Ing 77:318–320
Wu H, Li X, Berg P (2009) On the modeling of water transport in polymer electrolyte membrane fuel cells. Electrochim Acta 54(27):6913–6927
The OpenFOAM Foundation: OpenFOAM v6 User Guide. \url{https://cfd.direct/openfoam/user-guide}
Berg P, Kulikovsky AA (2015) A model for a crack or a delaminated region in a PEM fuel cell anode: analytical solutions. Electrochim Acta 174:424–429
Pasaogullari U, Wang CY (2004) Liquid water transport in gas diffusion layer of polymer electrolyte fuel cells. J Electrochem Soc 151(3):A399–A406
Beale SB et al (2009) Two-phase flow and mass transfer within the diffusion layer of a polymer electrolyte membrane fuel cell. Comput Therm Sci: Int J 1:105–120
Shi Y, Janßen H, Lehnert W (2019) A transient behavior study of polymer electrolyte fuel cells with cyclic current profiles. Energies 12(12)
Reimer U et al (2018) Irreversible losses in fuel cells. In: Hacker V, Mitsushima S (eds) Fuel cells and hydrogen - from fundamentals to applied research. Elsevier, pp 15–40
OpenFuelCell
Froning D et al (2016) Impact of compression on gas transport in non-woven gas diffusion layers of high temperature polymer electrolyte fuel cells. J Power Sources 318:26–34
Yu J et al (2018a) Liquid water breakthrough location distances on a gas diffusion layer of polymer electrolyte membrane fuel cells. J Power Sources 389:56–60
Yu J et al (2018b) Apparent contact angles of liquid water droplet breaking through a gas diffusion layer of polymer electrolyte membrane fuel cell. Int J Hydrogen Energy 43(12):6318–6330
Liang P et al (2018) Contact resistance prediction of proton exchange membrane fuel cell considering fabrication characteristics of metallic bipolar plates. Energy Convers Manage 169:334–344
Nam J et al (2010) Numerical analysis of gas crossover effects in polymer electrolyte fuel cells (PEFCs). Appl Energy 87:3699–3709
Chippar P, Ju H (2013) Numerical modeling and investigation of gas crossover effects in high temperature proton exchange membrane (PEM) fuel cells. Int J Hydrogen Energy 38:7704–7714
Chippar P et al (2014) Numerical analysis of effects of gas crossover through membrane pinholes in high-temperature proton exchange membrane fuel cells. Int J Hydrogen Energy 39:2863–2871
Kundu S et al (2006) Morphological features (defects) in fuel cell membrane electrode assemblies. J Power Sources 157(2):650–656
Karst N et al (2010) Improvement of water management in polymer electrolyte membrane fuel cell thanks to cathode cracks. J Power Sources 195(16):5228–5234
Kim S, Mench MM (2007) Physical degradation of membrane electrode assemblies undergoing freeze/thaw cycling: micro-structure effects. J Power Sources 174(1):206–220
Ding G et al (2016) Numerical evaluation of crack growth in polymer electrolyte fuel cell membranes based on plastically dissipated energy. J Power Sources 316:114–123
Kim S et al (2009) Investigation of the impact of interfacial delamination on polymer electrolyte fuel cell performance. J Electrochem Soc 156(1):B99–B108

# High Temperature Polymer Electrolyte Fuel Cell Model

**Qing Cao**

## 1 Introduction

Many categories of fuel cells (FCs) were developed over the past decade, based on different electrolytes and operating temperature. One promising type of FC is high-temperature polymer electrolyte fuel cell (HT-PEFC), it has a similar structure and functionality with the polymer electrolyte fuel cell (PEFC). The phosphoric acid-doped polymer is usually used as the electrolyte. The operating temperature of this device is between 150 and 180 °C. The HT-PEFC shows several advantages compare to PEFC, such as the relatively high carbon monoxide (CO) tolerance (1–2%) and non-flooding-water in gas channels and gas diffusion layers (GDL). A typical application of HT-PEFC is the fuel-cell-based auxiliary power units (APU) system. The high CO tolerance of HT-PEFC allows people to use reforming gas as the power source and gain a high net system efficiency of the APUs in the order of 20%, which is 2–12% higher than the conventional APUs.

The goal of HT-PEFC development is to develop the system with high performance, high durability and low cost. Therefore, a set of robust models on the cell level and system level are needed, to improve the understanding of the physics, and perform the prediction and optimization of HT-PEFC. The computational fluid dynamics (CFD) is a powerful technology for the modelling of the HT-PEFC. Many CFD modelling works are published based on commercial software such as STAR-CD, FLUENT, PHOENICS and COMSOL. On the other hand, the open-source-CFD

The original version of this chapter was revised: The ESM material has been updated. The correction to this chapter is available at https://doi.org/10.1007/978-3-030-92178-1_10

**Supplementary Information** The online version contains supplementary material available at https://doi.org/10.1007/978-3-030-92178-1_2.

Q. Cao (✉)
Forschungszentrum Jülich, Jülich, Germany
e-mail: caoqing85@hotmail.com

© Springer Nature Switzerland AG 2022, corrected publication 2022

S. Beale and W. Lehnert (eds.), *Electrochemical Cell Calculations with OpenFOAM*, Lecture Notes in Energy 42, https://doi.org/10.1007/978-3-030-92178-1_2

model (based on OpenFOAM) for FCs was developed and published by Beale et al. (2016; http://openfuelcell.sourceforge.net/project).

In this chapter you will learn how to use OpenFOAM to achieve the modeling of HT-PEFC. We will modify the OpenFuelCell library from Beale et al. to create the new model for HT-PEFC. As shown in the last chapter, the OpenFuelCell library solves the heat- and mass transfer equations in channels and porous zones for SOFC. It also provides useful classes, like calculation of multi-component diffusion coefficients, or heat capacity and dynamic viscosity of species. Those design patterns and classes can be further implemented in the HT-PEFC model. However, the major differences between HT-PFEC and SOFC should be noticed and treated carefully in the HT-PEFC model, they are:

- The composition of gas mixture: not like SOFC, the oxidant reduction reaction (ORR) of HT-PEFC happens at the cathode. It means the source term of water should be placed at the cathode catalyst layer. Typically, there are three species in the cathode side: $H_2O$ (gas), $O_2$ and $N_2$.
- The electrochemical kinetics: the activation loss and related parameters like the symmetry factor of the energy barrier and the exchange current density are different, based on the characteristics of the membrane, catalyst layer and operating conditions.
- The transfer of water: the membrane and electrodes of HT-PEFC can be considered as flooded with phosphoric acid. The production of water leads to concentration gradients, this in turn causes water diffusion in the acid. Additionally, the water content of membrane changes due to evaporation at the anode and cathode. The overall influence leads to a water transfer from one side to the other, a change in protonic resistance and a swelling of the membrane.

This chapter is organized as follow: first, the working principle, the components and working principle of HT-PEFC are presented. And then the OpenFOAM-model of HT-PEFC is introduced in two sections. In the first section, a basic model is introduced with C++ code, which incorporates the heat- and mass transfer equations and surface-related electrochemistry equations. The simulation results are compared with an analytical model: the Kulikovsky's model (Kulikovsky 2004). In the second section, an extended model of HT-PEFC is performed, to simulate the water crossover in the phosphoric acid flooded polymer membrane. The simulations are validated with the experimental results in Jülich and other publications.

## 2 Components and Principle of the HT-PEFC

The core part of HT-PEFC consists of Gas diffusion layers (GDLs), bipolar plates (BPPs) and a membrane electrode assembly (MEA). These components are shown in Fig. 1.

The HT-PEFC MEA is very different from other fuel cells: it has a poly(2,2′-mphenylene-5,5′-bibenzimidazole) (PBI) membrane doped with the concentrated

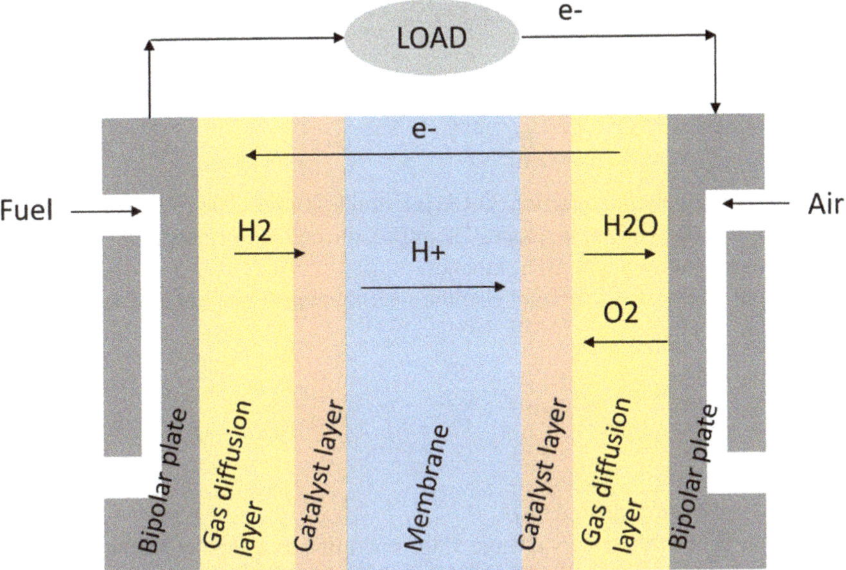


**Fig. 1** The components of the HT-PEFC

phosphoric acid and two catalyst layers (CLs) on anode and cathode side (Wainright et al. 1995). The acid doped membrane transfers protons from the anode to the cathode and prevent the passage of gas. That membrane has very good thermal stability, chemical stability and mechanical strength. However, the proton conductivity of the MEA is not constant: it depends on the weight percent (wt%) of phosphoric acid. Previous studies show that by 85 wt% of acid the conductivity reaches maximum. In the CL, the catalyst (Platinum) is mixed with supports (carbon blacks) and binders (polytetrafluoroethylene). The acid infiltrates into CL and the electrochemical reaction happens on the triple phase boundary of the mixture. The GDL has a porous solid structure, for allowing the reactant to move to the CL and the generated water to move to the gas channel. Typically, the GDL consists with carbon fiber paper with total thickness from 100 to 300 $\mu$m. The BPP provides the pathway for electrons and contains channels for allowing gas to pass through the flow field. The BPP is usually made by graphite or metal.

When the HT-PEFC starts working, the hydrogen and oxygen in the anode- and cathode side gas channels move through the GDLs and reacts in the CLs. The electrochemical reactions are:

Anode:

$$H_2 \rightarrow 2H^+ + 2e^- \tag{1}$$

Cathode:

$$\frac{1}{2}O_2 + 2H^+ + 2e^- \rightarrow H_2O(g) \tag{2}$$

Overall:

$$H_2 + \frac{1}{2}O_2 \rightarrow H_2O(g) \tag{3}$$

The oxygen reduction reaction (ORR) in cathode is much slower than the hydrogen oxidation reaction (HOR) in anode. Therefore, the cell performance of HT-PEFC is limited significantly by the ORR kinetics.

The cell open-circuit voltage, also the well-known Nernst voltage, is calculated as:

$$E_{Nernst} = \frac{-\Delta G^T}{2F} - \frac{RT}{2F}\ln\left(\frac{p_{H_2O}}{p_{H_2}p_{O_2}^{0.5}}\right) \tag{4}$$

where.

$E_{Nernst}$ = Nernst potential (V).

$\Delta G^T$ = Gibbs free energy change at the operating temperature (kJ $mol^{-1}$).

$p_{H2O}$, $p_{H2}$, $p_{O2}$ = normalized partial pressure ($p_i/p_0$) of $H_2O$, $H_2$ and $O_2$ (-).

In practice, the polarization curve is used to represent the cell performance. It is the plot of cell voltage against current density, as shown in Fig. 2. The cell voltage might be calculated as the difference between the Nernst voltage and the sum of other voltage losses.

$$E_{cell} = E_{Nernst} - \eta_{act} - \eta_{ohm} - \eta_{conc} \tag{5}$$

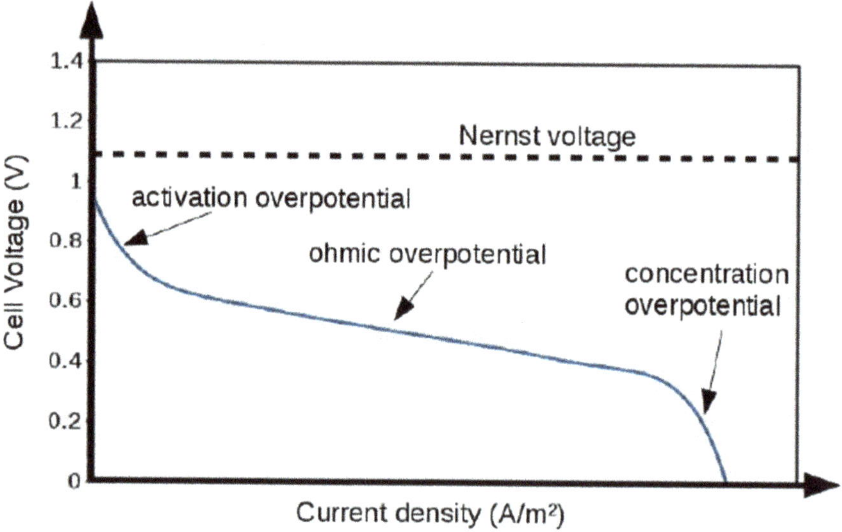


**Fig. 2** The polarization curve and all losses of HT-PEFC

where.
$E_{cell}$ = cell voltage (V).
$\eta_{act}$ = activation overpotential (activation loss) (V).
$\eta_{ohm}$ = ohmic overpotential (ohmic loss) (V) .
$\eta_{conc}$ = concentration overpotential (concentration loss) (V).

## 3 A Basic HT-PEFC OpenFOAM Model

In this section, we are going to create a simple HT-PEFC simulation application with OpenFOAM. The construction of this application contains two parts: create the mesh and create the solver. Let's talk about the mesh first.

### *3.1 Computational Domain and Meshing*

**Geometry**

The cross-section of the simulated physical domains of a HT-PEFC channel pair are shown in Fig. 4. Different governing equations are solved in different domains: the heat transfer is solved in 'Main domain'; the mass- and momentum transfer are solved in 'Domain I'; the chemical reactions are solved on the interface of 'Domain I' and 'Domain II'. The dimensions h1, h2, ... h5 are necessary parameters for creating the mesh, which are shown in Table 1. The typical thickness of CL is 100 μm which is relative thin compare to other components. We may assume 2-D CL surface and ignore the details of the triple phase boundary in this chapter, for reducing the complexity of meshing. Besides the channel pair, people have to notice that in practice the HT-PEFC system includes more hardware such as coolant supplies, current collectors and gaskets. They are not considered in this OpenFOAM model (Fig. 3).

**Mesh**

The next step is to generate the mesh. For simple flow fields, there are two options for generating meshes. One is to generate a 3D CAD model first, and then generate

**Table 1** 2-D drawing and parameters of HT-PFC cell structure

| Parameter | Symbol | Value | Unit |
|---|---|---|---|
| Height of gas channel | h1 | $1 \times 10^{-3}$ | m |
| Width of gas channel | h2 | $1 \times 10^{-3}$ | m |
| Thickness(total) of bipolar plate | h3 | $4 \times 10^{-3}$ | m |
| Thickness of GDL | h4 | $3 \times 10^{-4}$ | m |
| Thickness of membrane + CLs | h5 | $2 \times 10^{-4}$ | m |

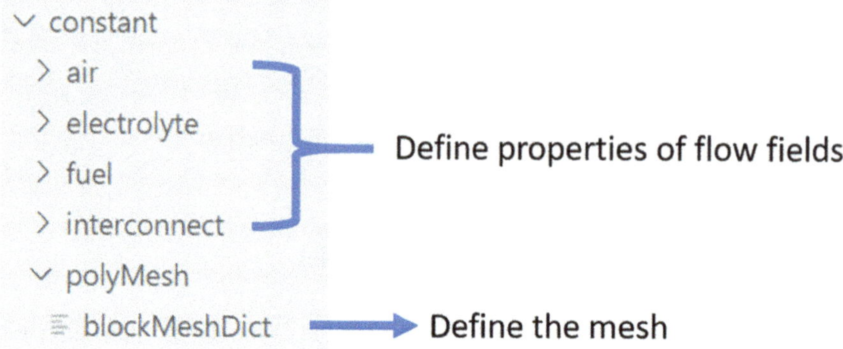


**Fig. 3** Code structure of the 'constant' dictionary

a mesh based on the CAD model. The other is to directly generate the mesh with the coordinate matrix using the OpenFOAM mesh-generation utility 'blockMesh'.

There are many options for the toolchain of the first path, such as SALOME Geometry module + SALOME Mesh module, ANSYS SpaceClaim + ANSYS ICEM CFD, CATIA V5 + PointWise, etc.

In this section we use the second path, the 'blockMesh'. First, we have to prepare a dictionary file named *blockMeshDict.* And put it in *constant/polyMesh* directory. The introduction of blockMesh can be found here (https://www.openfoam.com/documentation/user-guide/blockMesh.php).

The code structure of the 'constant' dictionary is shown as follow:

The mesh code in the *blockMeshDict* for a HT-PEFC single channel pair is showed as follow:

```
// ************************************************************ //
convertToMeters 0.001;
vertices
(
// Interconnect0
   (0 0 z0)
   (length 0 z0)
   (length width z0)
   (0 width z0)
// Interconnect0_to_Air
   (0 0 z1)
   (length 0 z1)
   (length width z1)
   (0 width z1)
// ************************************************************ //
```

As shown, we create a point coordinate list. In this list we define four points for a surface and eight points for a block. With this rule, the blocks for gas channels, BPPs, GDLs and the membrane of the HT-PEFC single-channel pair can be easily defined. In the remain part of the *blockMeshDict* file, the boundaries are defined by the index of the points:

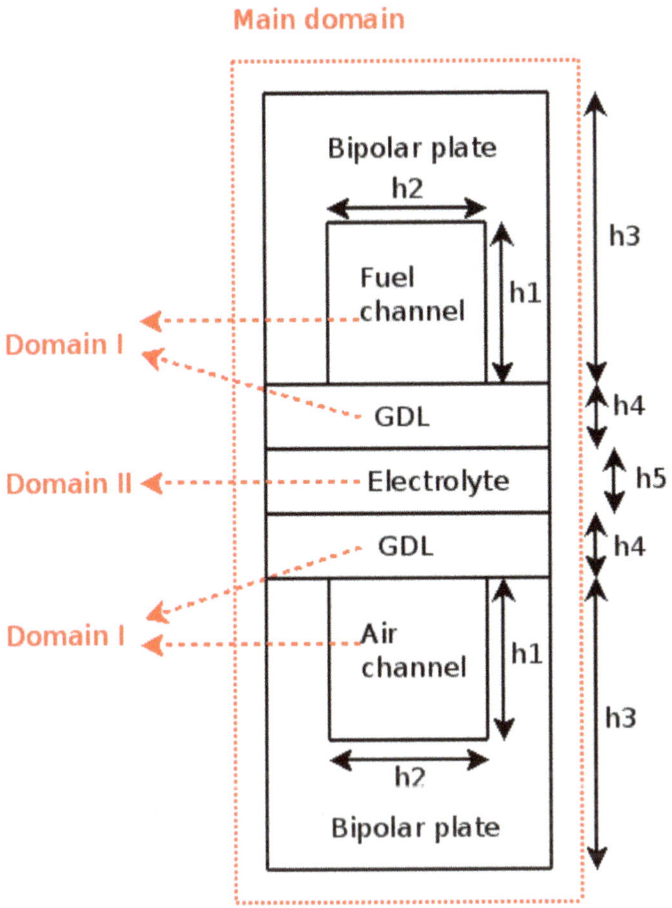

**Fig. 4** Cross-section of a single channel pair

```
// ********************************************************** //
 Patch airInlet
   (
      (4 8 11 7)
)
 Patch airOutlet
   (
      (5 6 10 9)
   )
...
// ********************************************************** //
```

After the mesh dictionary prepared, we can generate the mesh and boundaries with the command:

*blockMesh -case < case >*

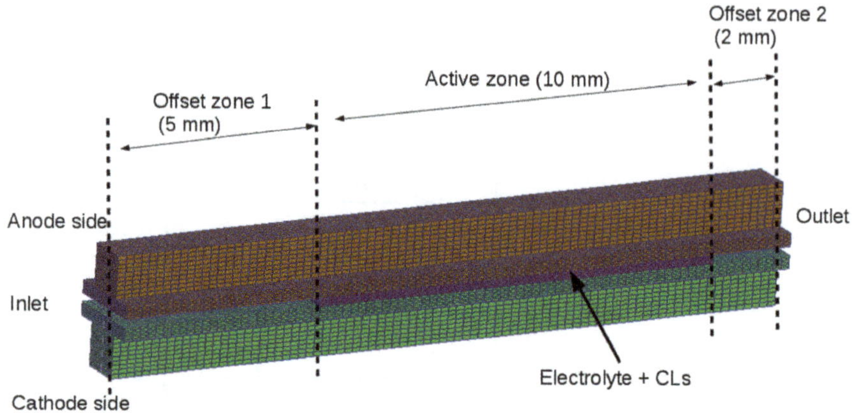


**Fig. 5** Mesh of the 17 mm straight single channel (BPP not shown). Brown mesh: anode side GDL and channel; green mesh: cathode side GDL and channel; red mesh: CLs and electrolyte

The mesh of the cell (without BPPs) is illustrated in Fig. 5, which is composed of 94,400 hexahedral blocks. Note that two offset zones are included in the mesh, to present the non-reactive gas channel zone of the inlet and outlet.

## 3.2 *Solver Design*

Now we have prepared the mesh, let's take a look at the solver. The first step is to design a code structure for this solver. We can use the code structure of the OpenFuelCell project (http://openfuelcell.sourceforge.net/project), as shown below (Fig. 6):

As shown, the project consists of two folders, *appSrc* and *libSrc*. *appSrc* stores code related to the specific fuel cell application, and *libSrc* stores code of general physical models. Readers can download the OpenFuelCell project to check the code.

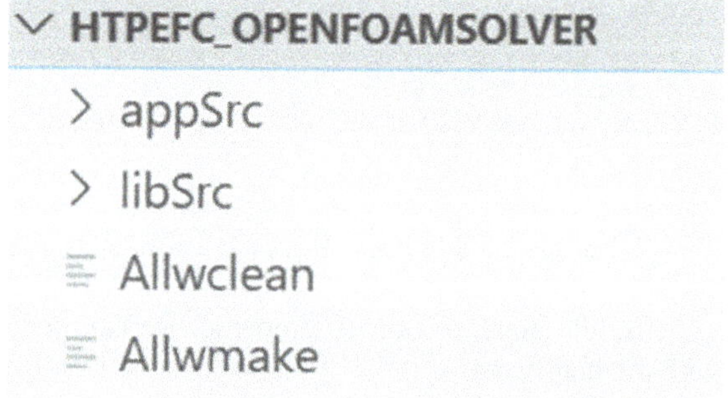


**Fig. 6** HTPEFC Solver structure

To create an HT-PEFC solver based on the OpenFuelCell project, people may do the following:

1. Add new model groups in *libSrc*, such as the electrochemical model group *'catalystLayerModels'*. That model group consists of one parent class and multiple child classes. For the simple HT-PEFC model, we create the electrochemical model file *'simpleNernst.C'*. The code structure is shown as follow (Fig. 7):
2. Modify all *'read_XXX_Properties.H'* files in *appSrc* to read the input of physical properties.
3. Modify all *'create_XXX_Fields.H'* files in *appSrc* to create and initialize the fields used by the solver.
4. Add *'createCatalystLayerModels.H'* in *appSrc* to create an instance of the electrochemical model.
5. Add *'htpefcFoam.C'* to *appSrc*, load all header files in a certain order, and complete the application.

The changes in the appSrc folder are shown in Fig. 8:

All fields should be declared as "create_XXX_Fields" and included by the main routine. A typical declaration looks like this:

```
// ************************************************************* //
  volScalarField Tair
  (
    IOobject
    (
      "T",
      runTime.timeName(),
      airMesh,
      IOobject::READ_IF_PRESENT,
      IOobject::AUTO_WRITE
    ),
```

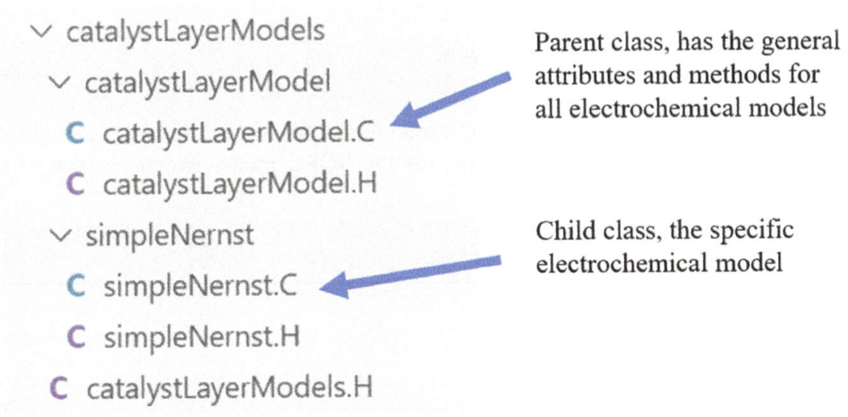

catalystLayerModels
catalystLayerModel
catalystLayerModel.C
catalystLayerModel.H
simpleNernst
simpleNernst.C
simpleNernst.H
catalystLayerModels.H
Parent class, has the general attributes and methods for all electrochemical models
Child class, the specific electrochemical model

**Fig. 7** Code structure of catalystLayerModels

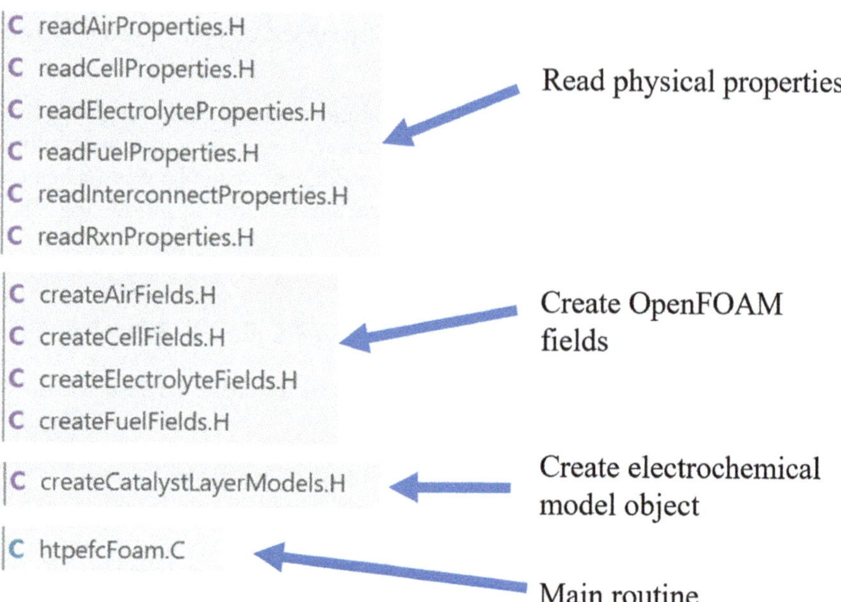


**Fig. 8** Changes in appSrc

```
        airMesh,
        Tinit,
        zeroGradientFvPatchScalarField::typeName
    );
// ************************************************************ //
```

In the code above, the `Tair` field is declared and installed on the mesh part `airMesh`. The electrochemical code is shown in the next section with the equations.

## 3.3 *Equations and Parameters*

The continuity equation, momentum transfer equation, energy transfer equation and the species transfer equation are same as the SOFC model which introduced in Chap. 1.

It is worth to note that the source term of produced water vapor is not applied in the anode like SOFC, but in the cathode:

$$S_i = \pm \frac{I}{nF} \tag{6}$$

where:

$S_i$ = species source term due to reaction (mol m$^{-2}$ s$^{-1}$).

I = local current density (A m$^{-2}$).

F = Faraday constant (C $mol^{-1}$).
n = electrons number (–).
The species is consumed or produced is defined in *runCase/constant/[field name]/ htpefcSpeciesProperties*. For anode side, the code:

```
// ************************************************************
H2    H2   2.01594 2 -1 0   130.680;
//    |    |       | |  |   |
//    |    |       | |  |   standardEntropy [J/(mol K)]
//    |    |       | |  enthalpyOfFormation [J/mol]
//    |    |       | produced=1|inert=0|consumed=-1
//    |    |       molecularChargeForFaradaysLaw
//    |    molecularWeight [kg/kmol]
//    name
// ************************************************************
```

For cathode side, the code:

```
// ************************************************************
O2    O2  31.9988   4  -1 0         205.152;
H2O   H2O 18.01534  2  1  -241.8349e3 188.835;
N2    N2  28.0134   0  0  0         191.609;
// ************************************************************
```

As shown, solver uses '1', '0' or '–1' to define the species is produced, inert or consumed in the cell. Number of electrons in Eq. 6 is also defined as '*molecularChargeForFaradaysLaw*'.

The biggest difference between HT-PEFC model and SOFC model is how to calculate the Nernst voltage and the activation overpotential. The Nernst voltage is given by:

$$E_{Nernst} = E_{th}^{T} - \frac{RT}{2F}\ln\left(\frac{X_{H_2O}}{X_{H_2}X_{O_2}^{0.5}}\right) \tag{7}$$

where:
$E_{Nernst}$ = Nernst potential (V).
$E_{th}{}^{T}$ = thermodynamic voltage at the cell operating temperature (V).
$X_{H2O}$, $X_{H2}$, $X_{O2}$ = mole fraction of species $H_2O$, $H_2$ and $O_2$ (–).
The code for Nernst voltage is in '*solver\appSrc\NernstEqn.H*':

```
// ************************************************************
Nernst =
relaxNernst.value()*(E0.value() - Rgas*anodeT*(Foam::log(XH2O)-
Foam::log(XH2)-0.5*Foam::log(XO2))/(2*F))          +          (1-
relaxNernst.value()) * Nernst;
// ************************************************************
```

Note that the `XH2,` `XH2O` and `XO2` in above code are field value, the calculated Nernst voltage is also a field. The under-relaxation is applied to stabilize the calculation by limiting the rate of change of fields.

The activation overpotential is given by:

$$\eta_{act} = \frac{RT}{\alpha F} \ln\left(\frac{I}{i_0}\frac{X_{ref}}{X_{O_2}}\right) \tag{8}$$

where:
$i_0$ = exchange current density for the ORR (A $m^{-2}$).
$\alpha$ = symmetry factor of the reaction (–).
$X_{ref}$ = reference mole fraction (–).
$X_{O2}$ = oxygen mole fraction (–).
The exchange current density is given by:

$$i_0 = i_0^{ref} \exp\left(-\frac{E_A}{R}\left(\frac{1}{T} - \frac{1}{T_{ref}}\right)\right) \tag{9}$$

where:
$i_0^{ref}$ = reference exchange current density for the ORR (A $m^{-2}$).
$E_A$ = activation energy for the ORR (J $mol^{-1}$).
$T_{ref}$ = reference temperature of operating (K).
The local current distribution is calculated by:

$$I = \frac{E_{Nernst} - E_{cell} - \eta_{act}}{R_\Omega} \tag{10}$$

where:
$E_{Nernst}$ = Nernst potential (V).
$\eta_{act}$ = activation loss (V).
I = current density of local electrolyte (A $m^{-2}$).
$R_\Omega$ = ohmic resistance (specific) ($\Omega$ $m^{-2}$).
The calculation of activation overpotential (field), exchange current density (single value) and current (field) are coded in *'solver\libSrc\catalystLayerModels\simpleNernst.C'*:

```
// ************************************************************
#include "simpleNernst.H"
namespace Foam
{
   namespace catalystLayerModels
   {
      defineTypeNameAndDebug(simpleNernst, 0);
      // Constructors
      simpleNernst::simpleNernst
      (
         scalarField& XO2,
         scalarField& YO2,
         scalarField& i,
         scalarField& eta,
         scalarField& Nernst,
         const dictionary& dict,
         dimensionedScalar& Temp,
         dimensionedScalar& Vcell
      )
      :
```

```
        catalystLayerModel(XO2, YO2, i, eta, Nernst, dict),
        j0Ref_(dict_.lookup("j0Ref")),
        Tref_(dict_.lookup("Tref")),
        alpha_(dict_.lookup("alpha")),
        Rohm_(dict_.lookup("Rohm")),
        F_(dict_.lookup("F")),
        Rgas_(dict_.lookup("Rgas")),
        EA_(dict_.lookup("EA")),
        relax_(dict_.lookup("relax")),
        XO2ref_(dict_.lookup("XO2ref")),
        Temp_(Temp),
        Vcell_(Vcell)
        {}
        // Members functions
        void simpleNernst::evaluate()
        {
        scalar kt = (Rgas_.value() * Temp_.value()) / (F_.value() *  \\
   alpha_.value());
   scalar j0 = j0Ref_.value()* Foam::exp( (-EA_.value()
   / Rgas_.value()) * (1.0 / Temp_.value() - 1.0 / Tref_.value() ) );
            eta_ = kt * Foam::log(i_/(j0 * (XO2_/XO2ref_.value())));
            i_  = relax_.value()*(Nernst_ - Vcell_.value() - eta_)/
    Rohm_.value() + (1-relax_.value())* i_;
        }
     }
   }
   // ************************************************************
```

Equations 8–10 are solved in the members function `simpleNernst::evaluate()`. The under-relaxation is used for calculating the current. Note that in the constructor, the necessary `scalarField` such as `XO2`, `i` or `Nernst` has to be declared and pointed to an initialized `scalarField` with pointer symbol &. The physical constants are read with the `dict_.lookup()` function. The `simpleNernst` model object itself is created with *solver\appSrc\ createCatalystLayerModels.H:*

```
   // ************************************************************
   ...
           const  dictionary&  simpleNernstDict  =  cellProper-
   ties.subDict("simpleNernst");
     catalystLayerModels::simpleNernst simpleNernst = catalystLay-
   erModels::simpleNernst(XO2, YO2, j_O2CL, eta, Nernst, simpleNern-
   stDict, Temperatur, V);
   ...
   // ************************************************************
```

The heat capacity and dynamic viscosity of species by typical operating temperature of HT-PEFC are calculated using 6th order polynomial according to Todd and Young's research. The other modelling parameters are shown in Table 2. Those parameters are assumed to be constant and easily to be implemented to the code. By default, OpenFOAM uses the SI units with 7 scalars delimited by square brackets. The parameters are defined in '*runCase/constant/cellProperties*':

**Table 2** Modelling parameters of the basic model

| Parameter | Symbol | Value | Unit |
|---|---|---|---|
| Symmetry factor | $\alpha$ | 0.5 | |
| Ohmic resistance of cell per m$^2$ | $R_\Omega$ | $2 \times 10^{-5}$ | $\Omega m^{-1}$ |
| Reference exchange current density | $i_0^{ref}$ | 3.0 | $Am^{-2}$ |
| Reference mole fraction of oxygen | $X_{ref}$ | 0.23 | |
| Reference temperature | $T_{ref}$ | 433 | K |
| Thermodynamic voltage | $E_{th}^T$ | 1.15 | V |
| Activation energy | $E_A$ | 57100 | $Jmol^{-1}$ |
| Thickness of membrane + CLs | $l_e$ | $2 \times 10^{-4}$ | m |

```
// *************************************************************
...
alphaalpha[0 0 0 0 0 0 0] 0.5;
Rohm Rohm [1 4 -300 -20] 2.0e-5;
...
// *************************************************************
```

Note that by other cell assembly the parameters should be different, due to the different porous GDL, composite of catalyst layer, situation of membrane, etc.

## *3.4 Results*

**Polarisation curve**

In this chapter, eight operating conditions with average of the local current densities from 1000 A/m$^2$ to 8000 A/m$^2$ were used to calculate the operating range of an HT-PEFC.

The operating temperature was 433 K, the gas pressure at both anode and cathode side are 101325 Pa, the mass fraction of species is: 100% $H_2$ at anode, 23% $O_2$ and 77% $N_2$ at cathode. Stoichiometric ratio at both sides is 2.

For each operating current, the cell voltage was calculated by the solver according to Eq. 10. A polarization curve was therefore plotted, as shown in Fig. 9. People can find that the polarization curve of CFD-simulations fits the in-house experiments well.

**Local distribution**

Since the cathode side species distribution determines the performance of the HT-PFEC, we will use cathode side in this chapter to show the numerical results. In the 3D flow field of the cathode, we choose two section-plane to show the distribution of the species: plane A and B. In Fig. 10. we put the plane A in the center of the channel and put plane B on the electrolyte-GDE interface.

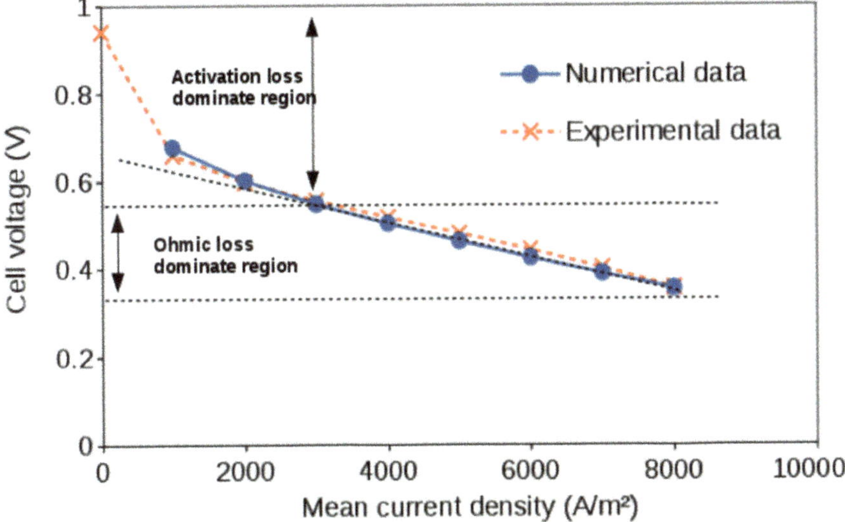


**Fig. 9** Comparison of numerical and the experimental results of the polarisation curve

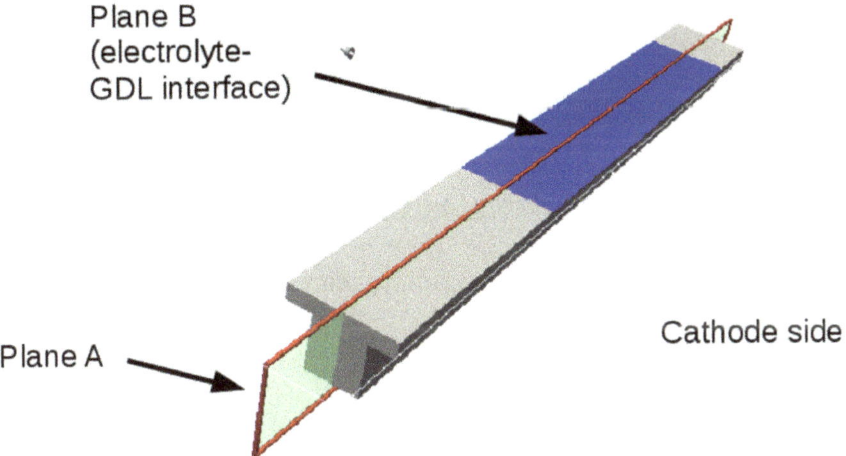


**Fig. 10** The position of the cross-sectional plane A and B. To present the local simulation results

We may choose 2 operating points to compare the species distribution at the low- and high current density: 2000 and 8000 A $m^{-2}$. The oxygen mole fraction on plane A and plane B of these 2 sampling points are shown in Figs. 11 and 12.

For HT-PEFC, there is an analytical model by A. Kulikovsky (2004) to calculate the mole fraction of oxygen in the channel. Note that this equation assumed plug flow.

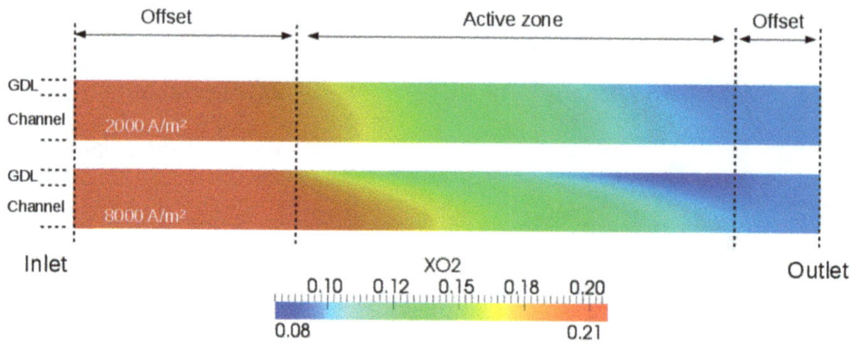


**Fig. 11** Local distribution of oxygen mole fraction at 2000 A $m^{-2}$ and 8000 A $m^{-2}$ on Plane A

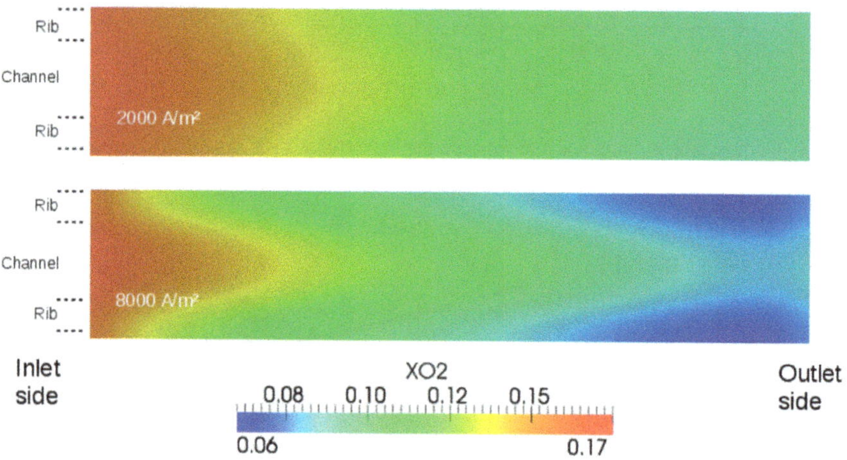


**Fig. 12** Local distribution of oxygen mole fraction at 2000 A $m^{-2}$ and 8000 A $m^{-2}$ on Plane B

$$X_{O_2}(x) = X_{O_2}^0\left(1 - \tfrac{1}{\lambda}\right)^{\frac{x}{x0}} \tag{11}$$

where:

$X_{O2}(x)$ = oxygen mole fractions along the channel direction.

$X_{O2}{}^0$ = oxygen mole fractions at the input.

λ = stoichiometric ratio of HT-PEFC.

x = distance (m).

$x_0$ = length of the reaction zone (m).

This model can be used to verify our numerical results. Use the same operating parameters as the CFD, we can get the analytical curve of $X_{O2}$. The difference between numerical and analytical calculation is shown in Fig. 13. People should note that the $X_{O2}$ in the Kulikovsky model is a singel value and in the CFD model is a field value. In Fig. 13 we averaged the CFD result of $X_{O2}$ to compare it with the Kulikovsky model.

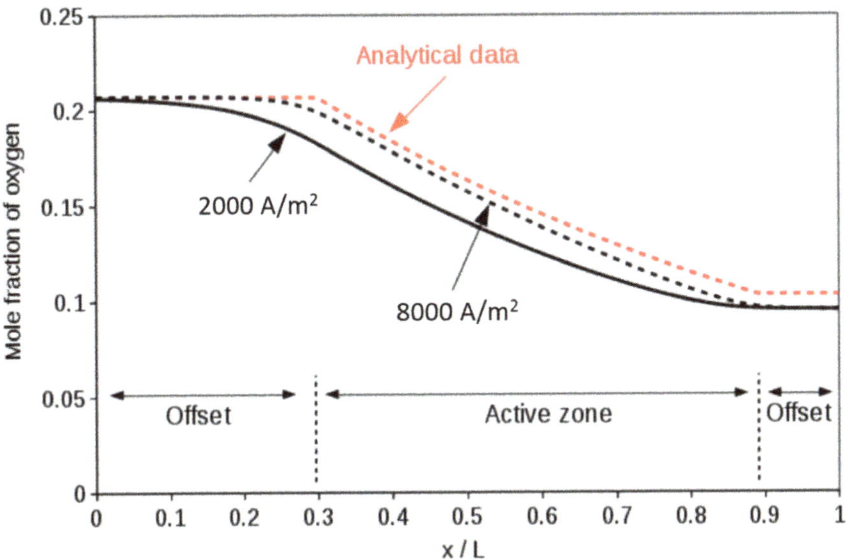


**Fig. 13** Comparison of oxygen mole fraction distribution along the channel at 2000 A $m^{-2}$ and 8000 A $m^{-2}$ on Plane A

People can see that the CFD-simulation and the analytical calculation have similar trends: the mole fractions of oxygen decrease along the gas channel direction. However, the CFD result at high current fit better with the analytical results than low current. The reason is that the Péclet number at low current is much smaller than high current, so the diffusion effect becomes obvious.

## 4 An Extended HT-PEFC OpenFOAM Model

The previous section we describe how to build a simple HT-PEFC modeling project. In this chapter, we will show how to add some advanced modules to the basic model.

In HT-PEFC, the water balance affects the cell performance significantly, because (i) in the membrane, the proton conductivity depends significantly on the phosphoric acid concentration (Wainright et al. 1995) and (ii) the partial pressure of water vapor on electrode affects the kinetics. The Fig. 14. shows the water transport process during cell operation: water is first produced at the membrane-GDE interface of the cathode, and then water vapor is absorbed into the membrane, resulting in a difference in water concentration on both sides of the membrane. The water flows from the cathode side to the anode side due to the chemical potential, and finally enters the GDL from the anode side. The process of water entering and leaving the membrane is controlled by the evaporation-adsorption process.

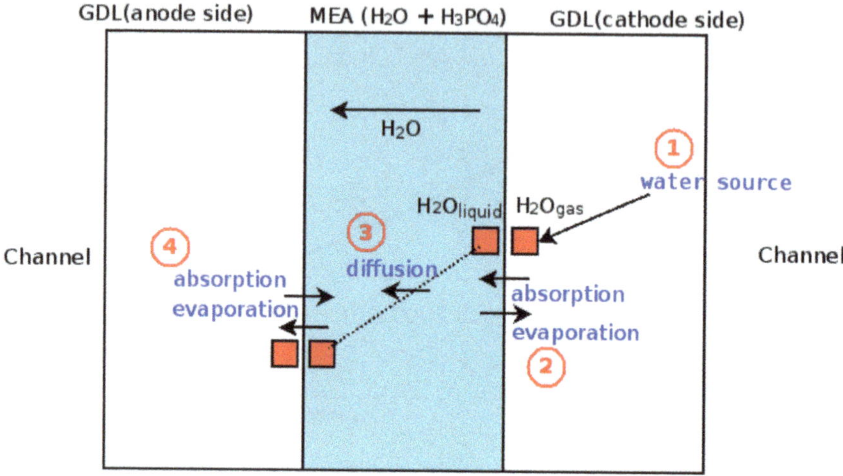


**Fig. 14** Schematic model of water transfer in electrolyte

## *4.1 Computational Domain and Meshing*

**Geometry**

In this section we will create the mesh corresponds to an existing HT-PEFC cell (Fig. 15.) in Jülich. The active area is around 50 cm$^2$. The dimension of cross-section of the simulated physical domains are the same as last section.

**Mesh**

We cannot use the 'blockMesh' for this project due to the complexity of the geometry. For such complex shapes, we can use the open-source pre-processing tool ‚Salome'.

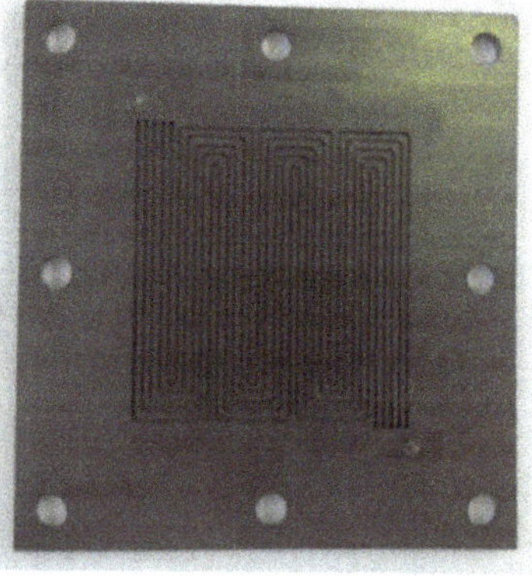

**Fig. 15** *Left: 5-channel serpentine test cell in Jülich; right: the bipolar plate of this cell* (Cao 2017)

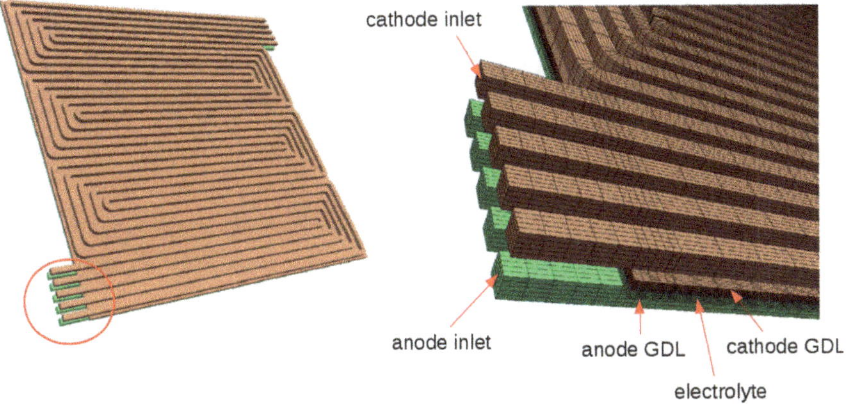


**Fig. 16** Left: 3-D geometry of the 5-channel serpentine test cell (BPP not shown); right: the mesh of inlet area

There is truly marvelous meshing process with Salome, which this margin is too narrow to contain. The introduction of Salome Meshing can be found here (https://docs.salome-platform.org/7/gui/SMESH/index.html). The mesh of this single cell is illustrated in Fig. 16, which composed of 0.74 million hexahedral volumes.

## 4.2 *Solver Design*

We have already completed the project architecture in the previous chapter. Adding new modules base on that becomes simple:

1. Add water transfer model groups 'waterTransferModels' in libSrc, create parent class 'waterTransferModel' add child class 'diffusionDrive'. The code structure is shown as follow (Fig. 17):
2. Modify all 'read_XXX_Properties.H' files in appSrc to read the input of physical properties of new module.
3. Modify all 'create_XXX_Fields.H' files in appSrc to create and initialize the fields of new module.
4. Add 'createWaterTransferModels.H' in appSrc to create an instance of the water transfer model.
5. Modify 'htpefcFoam.C' in appSrc, load all header files in a certain order, and complete the application.

The specific code will be shown in the next section based on the water transfer equations.

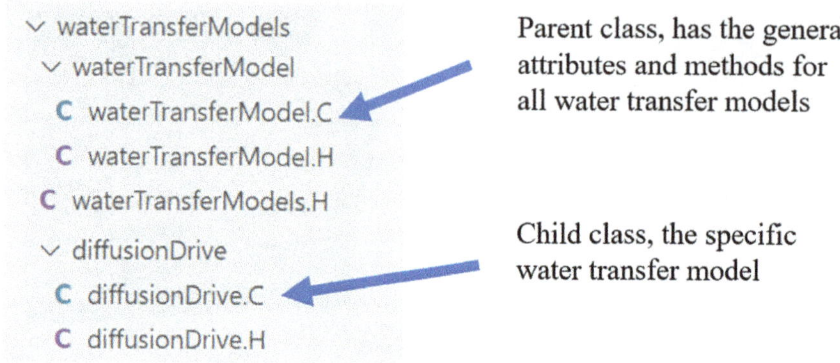


**Fig. 17** Code structure of waterTransferModels

## 4.3 Equations and Parameters

In the $H_3PO_4$ electrolyte, the diffusion flux of the water a function of the gradient of the liquid water mass fraction. That diffusion flux is given by:

$$j_{H_2O} = \frac{\psi k_{H_2O,l}^{eff}\left(Y_{H_2O}^{liq.c} - Y_{H_2O}^{liq.a}\right)}{l_e} \tag{12}$$

where:
$\psi$ = empirical parameter of the effect of mean current (-).
$Y_{H2O}^{liq.c}$ = liquid water mass fraction at cathode (-).
$Y_{H2O}^{liq.a}$ = liquid water mass fraction at anode (-).
$l_e$ = thickness of electrolyte (m).

Note that we introduce an empirical factor $\psi$ here. Because by different mean current densities, the water-acid solution density also changes. Therefore in the electrolyte, the effective mass transfer coefficient of water might be calculated as a function of current density:

$$\psi = \frac{i_m}{i_{ref,w}} + 0.75 \tag{13}$$

where:
$i_m$ = mean current density (A $m^2$).
$i_{ref,w}$ = reference current density (A $m^2$).

The boundary values of liquid water mass fraction are governed by water vapor partial pressure on the gas–liquid interface. Using the research from Schechter et al. (2009), we may get an empirical equation by fitting of their experimental data:

$$Y_{H_2O}^{liq,b} = 0.517\frac{p_{H_2O}}{p_{sat}} + 0.046 \tag{14}$$

where:

$Y_{liq,b}$ = liquid water mass fraction on the gas–liquid interface (-).
$p_{H2O}$ = water vapor partial pressure (Pa).
$p_{sat}$ = saturation water vapor pressure (Pa).

The value of the saturation vapor pressure of water is calculated by Antoine equation (an Clausius-Clapeyron equilibrium formulation):

$$p_{sat} = 133.322 * 10^{a - \frac{b}{c + \tau_0 - 273.15}} \tag{15}$$

where a = 8.14, b = 1810.94, c = 244.49 are empirical factors from an online database: Dortmund Data Bank (http://ddbonline.ddbst.de/AntoineCalculation/AntoineCalculationCGI.exe), tau is the temperature.

Other modelling parameters of this model are listed in Table 3.

Since the saturation vapor pressure of water (Eq. 15) and activity of vapor $p_{H2O}$ / $p_{sat}$ are general fields that can be used for all water transfer models, they should be solved in the parent class *'solver\libSrc\waterTransferModels\waterTransferModel\waterTransferModel.C'*:

```
// ************************************************************
...
    // Member Functions
    //- solve the saturated vapor pressure of water in H3PO4
    void waterTransferModel::solvePsat()
    {
      // solve partial pressure of water
      Info << nl << "Solving partial pressure" << endl;
      pPAir_= pAir_*XH2Ocathode_;
      pPFuel_= pFuel_*XH2Oanode_;
      // solve saturation vapor pressure for pure H2O, DDBST
    Info << nl << "Solving water saturation vapor pressure" << endl;
                              pWSAir_   =   pow(10,   AEP_A_.value()-
AEP_B_.value()/(AEP_C_.value()+Tair_ -273.15) ) * 133.322;
                              pWSFuel_=   pow(10,   AEP_A_.value()-
AEP_B_.value()/(AEP_C_.value()+Tfuel_ -273.15) ) * 133.322;
      // solve water activity
      Info << nl << "Solving water activity" << endl;
      aWAir_ = pPAir_/pWSAir_;
      aWFuel_ = pPFuel_/pWSFuel_;
    }
    void waterTransferModel::relaxGamma()
    {
                    gamma_  =   relax_.value()  *  gamma_  +  (1-
relax_.value()) * gammaOld_;
```

**Table 3** Modelling parameters in the water transfer model

| Parameter | Symbol | Value | Unit |
|---|---|---|---|
| Effective mass transfer coefficient | $k_{H2O,l}^{eff}$ | $2 \times 10^{-6}$(assumed) | $kgm^{-1}s^{-1}$ |
| Reference current density | $i_{ref,w}$ | $1 \times 10^{4}$(assumed) | $Am^{-2}$ |

```
        infoScalarField(gammaOld_, "gammaOld");
    }
...
// ************************************************************
```

Equations 12–14 are solved in the child class *'solver\libSrc\waterTransferModels\diffusionDrive\diffusionDrive.C'*:

```
// ************************************************************
#include "diffusionDrive.H"
#include "volFields.H"
namespace Foam
{
  namespace waterTransferModels
  {
    defineTypeNameAndDebug(diffusionDrive, 0);
    // Constructors
    diffusionDrive::diffusionDrive
    (
      scalarField& XH2Oanode,
      scalarField& XH2Ocathode,
      scalarField& gamma,
      const dictionary& dict,
      scalarField& YwaterC,
      scalarField& YwaterA,
      scalarField& prodFluxWater,
      scalarField& pAir,
      scalarField& pFuel,
      scalarField& Tair,
      scalarField& Tfuel,
      scalarField& Telectrolyte
    )
    :
   waterTransferModel(XH2Oanode, XH2Ocathode, gamma, dict, YwaterC,
\\ YwaterA, pAir, pFuel, Tair, Tfuel, Telectrolyte),
    prodFluxWater_(prodFluxWater),
    lMem_(dict_.lookup("lMem")),
    Res_(dict_.lookup("Res")),
    kWater_(dict_.lookup("kWater"))
    {}
    // Member Functions
    void diffusionDrive::evaluate()
    {
      writeData();
      gammaOld_ = gamma_;
      solvePsat();
                                          //  solve  water  mass  frac-
tion boundary in membran, fitting from Schechter et al.
      YwaterC_ = 1-(-51.7 * aWAir_ +95.4)/100;
      YwaterA_ = 1-(-51.7 * aWFuel_ +95.4)/100;
      // solve water flux through membrane
        fluxWater_ = Res_.value()*kWater_.value() * (YwaterC_-
YwaterA_) / lMem_.value();
      // update gammaWater
```

```
            gamma_= fluxWater_ / prodFluxWater_;
            relaxGamma();
        }
    }
}
// ************************************************************************
```

## 4.4 Results

### Water outcomes on electrodes

In this section, the water transfer models are validated with the experiment: an in-house HT-PEFC cell was operated with the same conditions as the simulations (Reimer et al. 2016). The water was condensed by two condensers at outlets of anode and cathode and then collected using bottles. Figure 18. shows the comparison of the numerical results and the experiment results. One can see that the water outcomes and the water crossover rate of the CFD simulations fit the experiments well.

### Local distribution

The CFD model calculated the scalar field of water and $H_3PO_4$ in the electrolyte of HT-PEFC. The sum of the mass fractions of water and $H_3PO_4$ equals 1. Figure 19. shows the mass fractions of $H_3PO_4$ at the electrolyte-GDE interface. One can see that the $H_3PO_4$ concentration of the anode is higher than cathode. That concentration

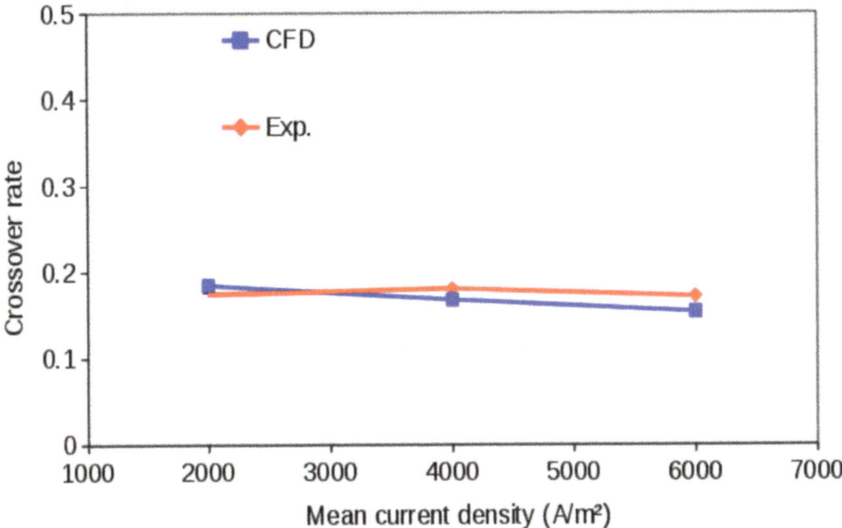


**Fig. 18** Cross over rate of the produced water: comparison of CFD and experiment (Cao 2017)

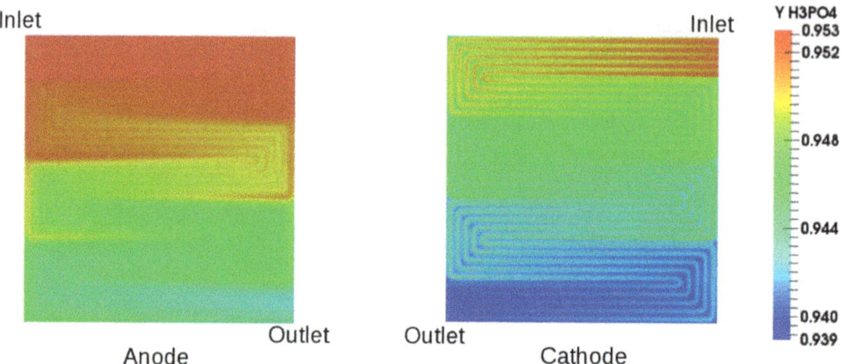


**Fig. 19** $H_3PO_4$ Mass fraction in the electrolyte: comparison of anode and cathode

gradient is the driving force of the water cross over. In the membrane of HT-PEFC, the CFD calculation indicates that mass fractions of $H_3PO_4$ is between 93.9% and 95.3%. This result is validated with a previous experiment, which applied synchrotron-based X-ray tomographic microscopy to an HT-PEFC single cell: the mass fractions of $H_3PO_4$ is reported as $96.5 \pm 1.5$ wt % at 160 °C (Eberhardt et al. 2015).

## References

Beale SB, Choi H-W, Pharoah JG, Roth HK, Jasak H, Jeon DH (2016) Open- source computational model of a solid oxide fuel cell. Comput Phys Commun 200:15–26

Cao Q (2017) Modelling of high temperature polymer electrolyte fuel cells. Universitätsbibliothek der RWTH Aachen, Diss

Eberhardt S, Toulec M, Marone F, Stampanoni M, Büchi F, Schmidt T (2015) Dynamic operation of HT-PEFC: in-operando imaging of phosphoric acid profiles and (Re) distribution. J Electrochem Soc 162(3):F310–F316

Kulikovsky A (2004) The effect of stoichiometric ratio on the performance of a polymer electrolyte fuel cell. Electrochim Acta 49(4):617–625

Mesh generation with the blockMesh utility, https://www.openfoam.com/documentation/user-guide/blockMesh.php

Reimer U, Ehlert J, Janssen H, Lehnert W (2016) Water distribution in high temperature polymer electrolyte fuel cells. Int J Hydrogen Energy 41(3):1837–1845

Salome Mesh User's Guide, https://docs.salome-platform.org/7/gui/SMESH/index.html

Saturated Vapor Pressure Calculation by Antoine Equation, http://ddbonline.ddbst.de/AntoineCalculation/AntoineCalculationCGI.exe

Schechter A, Savinell RF, Wainright JS, Ray D (2009) 1H and 31 P NMR study of phosphoric acid-doped polybenzimidazole under controlled water activity. J Electrochem Soc 156(2):B283–B290

The openFuelCell project, http://openfuelcell.sourceforge.net/project

Wainright J, Wang J, Weng D, Savinell R, Litt M (1995) Acid-doped polybenzimidazoles: a new polymer electrolyte. J Electrochem Soc 142(7):L121–L123

# Solid Oxide Fuel Cell Modeling with OpenFOAM®

**Dong Hyup Jeon**

## Nomenclature

| | |
|---|---|
| $Cp$ | heat capacity, J/kg K |
| $D$ | diffusivity, $m^2/s$ |
| $d_p$ | diameter of spherical particle, m |
| $F$ | Faraday constant, 96,485 C/mol |
| $\Delta G^{\circ}$ | gibb's free energy, J/kmol |
| $H$ | enthalpy, J/kmol |
| $i$ | current density, $A/m^2$ |
| $K$ | flow permeability, $m^2$ |
| $l$ | thickness, $\mu m$ |
| $M_i$ | molecular mass of gas species $i$, kg/mol |
| $M_{ij}$ | mean molecular mass, kg/kmol |
| $P$ | pressure, Pa |
| $R_g$ | universal gas constant, 8.314 J/K mol |
| T | temperature, K or °C |
| $u$ | velocity, m/s |
| $V$ | voltage, V |
| $V_i$ | diffusion volume of species $i$ |
| $x_i$ | mole fraction of component gas species $i$ |
| $y_i$ | mass fraction of component gas species $i$ |
| | Greek Letters |
| $\varepsilon$ | Porosity |
| $\eta$ | overpotential, V |
| $\kappa$ | thermal conductivity, W/m K |

D. H. Jeon (✉)
Department of Mechanical System Engineering, Dongguk University-Gyeongju, Gyeongju 38066, Republic of Korea
e-mail: jeondh@dongguk.ac.kr

© Springer Nature Switzerland AG 2022

S. Beale and W. Lehnert (eds.), *Electrochemical Cell Calculations with OpenFOAM*, Lecture Notes in Energy 42, https://doi.org/10.1007/978-3-030-92178-1_4

| | |
|---|---|
| $\mu$ | dynamic viscosity, Pa s |
| $\rho$ | density, kg/m$^3$ |
| $\tau$ | tortuosity |
| | Superscripts or Subscripts |
| 0 | initial |
| A | anode |
| act | activation |
| C | cathode |
| conc | concentration |
| eff | effective |
| $H_2$ | hydrogen |
| $H_2O$ | water |
| $O_2$ | Oxygen |
| T | total |

## 1 Introduction

Solid oxide fuel cells (SOFCs) are an electrochemical device that converts the chemical energy stored in a variety of fuels directly into electricity and produces heat as its by-product. SOFCs typically operate in high temperature range (600–1000 °C) to achieve a good ionic conductivity of solid electrolyte. High temperature operation has advantages, such as high power efficiency, long-term stability, fuel flexibility and high temperature exhaust gas. This benefit further allows the system to be combined with gas turbine that can be an option for applications of auxiliary power unit of vehicles as well as industrial power supply. One the other hand, disadvantages include longer start-up time, thermal fatigue failure of the materials, sealing problem under high temperature, and mechanical and chemical compatibility issues. Therefore, the SOFCs operation within a tight design range is important (Hajimolana et al. 2011). In this respect, numerical modeling has become an attractive method for investigating the SOFCs that can optimize the cell parameters, decrease physical and chemical degradation, and finally improve its performance.

There are two principal SOFC configurations, e.g., tubular and planar. Typically, planar configuration consists of positive electrode/electrolyte/negative electrode (PEN) structure, gas channels and interconnects. Based on the component thickness, SOFCs are categorized to four types, e.g., anode-supported, electrolyte-supported, cathode-supported and metal-supported. When operating SOFCs, a fuel and an oxidant, supplied from gas channels, are introduced to the PEN structure. Both reactant gases are electrochemically consumed, and generate the water and current. In an actual operation, SOFCs are connected in series of multiple cells to increase voltage, thereby obtaining a high output power. An interconnect functions as electrical connectors when the current flows through the cell. The SOFC operation involves the simultaneous transport of mass, momentum, species and heat which are fully coupled with electrochemical reactions. To comprehend those processes and assist the development of SOFCs technology, numerical simulations based on computational fluid dynamics (CFD) approach are widely used in SOFC research by exploiting the transport phenomena and electrochemical process. CFD simulation is varied upon the research's concerns. Micro-scale model pursues to understand the effect of microstructural properties on the electrochemical characteristics and transport phenomena (Bertei et al. 2013; Costamagna et al. 1998; Chan and Xia 2001; Zhu et al. 2005; Nam and Jeon 2006; Hussain et al. 2006; Kong et al. 2014; Vijay et al. 2017). However, this model is incapable of providing the information in cell-level which is useful in practical application. Macro-scale model deals with physicochemical hydrodynamics in cell- and stack-level and predicts the cell performance with experimentally determined model parameters, but does not pursue the performance prediction in consideration of various microstructures (Achenbach 1994; Nagata et al. 2001; Yakabe et al. 2001; Recknagle et al. 2003; Pramuanjaroenkij et al. 2008; Ni 2012). Recent CFD modeling focuses the development of multi-scale model which in principle combines the advantages of both micro-scale and macro-scale models. Here, we employ the multi-physics and multi-scale model to simulate the SOFC operation.

Recently, the cell-level SOFC model on the open-source software is developed by the author and co-workers (Beale et al. 2016; Jeon 2019). The work was originated from the Multi-Scale Integrated Fuel Cell Model (MUSIC) program, in which the goal is to develop a fully-integrated multi-scale fuel cell modeling capability, from micro-scale through to cell, stack, and 'hotbox' levels. We select the object-oriented CFD code, Open Field Operation and Manipulation (OpenFOAM) (http://www.openfoam.com), as a development platform and released the source code on the internet (http://openfuelcell.sourceforge.net). The public availability of open-source fuel cell code, openFuelCell, has stimulated the activities of code development (Beale et al. 2011, 2018; Nishida et al. 2018) and application to other types of fuel cell (Kone et al. 2018; Zhang et al. 2018). The openFuelCell is a multi-physics and multi-scale model that the users easily access the source code and free to modify the code as desired. The model describes complex transport and electrochemical phenomena in a variety of geometrical configurations. This model enables to deal with multiple regions within a single solver, with efficient data transfer between them. Each region supports its own set of fields such as pressure, velocity, concentration and temperature, with associated boundary conditions, while computational domain is decomposed into several regions. Validation and verification activities have been carried out under the program of the International Energy Agency Advanced Fuel Cells Implementing Agreement (IEA AFCIA) (Nishida et al. 2018; Le et al. 2015). However SOFC modeling with openFuelCell has several challenges, such as limited application to planar configuration, unavailability of internal reforming process, high-calculation cost, etc. The present software extends the capability to simulate the stack model.

In this chapter, we introduce the open-source SOFC model which is developed with OpenFOAM®. The widely used anode-supported SOFC is chosen to describe the modeling process. The details of SOFC modeling are described in each section: model equations in Sect. 2; numerical implementation including calculation procedure in Sect. 3; model description including geometry, mesh and boundary condition in Sect. 4; results and discussion in Sect. 5.

# 2 Model Equations

## *2.1 Governing Equations*

CFD model simulates the heat and mass transfer phenomena in SOFCs. The multi-scale and complex transport phenomena are based on the coupled set of conservation equations of mass, momentum, species transport and energy. The conservation of mass can be expressed in the form of the steady-state equation of continuity:

$$\mathrm{div}\left(\rho \vec{u}\right) = S_m \tag{1}$$

$$S_m = \sum_{i=1}^{n} S_i \tag{2}$$

where the source terms are in general non-zero as a result of production of water and consumption of hydrogen and oxygen caused by electrochemical reactions. The conservation of linear momentum for steady-state flow in the absence of external body forces yields:

$$\mathrm{div}\left(\rho \vec{u} \cdot \vec{u}\right) = -\mathrm{grad}P + \mathrm{div}\left(\mu \mathrm{grad}\vec{u}\right) + S_p \tag{3}$$

$$S_p = -\frac{\mu \vec{u}}{K} \tag{4}$$

where the source term on the right hand side of the equation is only added in the porous regions such that neglecting the convection and viscous terms in Eq. (3) yields conservation of linear momentum expressed in the form of Darcy's law. Using the generalized Fick's law and neglecting the thermal-diffusion (Soret) effect, the conservation of species mass yields:

$$\mathrm{div}\left(\rho \vec{u} y_i\right) = \mathrm{div}\left(D_i^{eff} \mathrm{grad} y_i\right) + S_i \tag{5}$$

$$S_{H_2} = -\frac{i}{2F} M_{H_2},\ S_{H_2O} = \frac{i}{2F} M_{H_2O},\ S_{O_2} = -\frac{i}{4F} M_{O_2} \tag{6}$$

where the effective Fick diffusion coefficients are given in terms of values at reference pressure and temperature. Neglecting kinetic energy convection, viscous dissipation, and the diffusion–thermo (Dufour) effect, conservation of energy for steady-state flow yields:

$$\mathrm{div}\left(\rho C_p \vec{u}\, T\right) - \mathrm{div}(k\,\mathrm{gradT}) = S_h^e + S_h^j \tag{7}$$

$$S_h^e = \frac{i}{2F}\Delta H(T) - i V^{cell},\ S_h^j = i^2 R \tag{8}$$

where the source term takes into account local heat production due to joule heating and reversible enthalpy changes. More detailed description for the equations can be found in Beale et al. (2016) and Jeon (2019).

## 2.2 *Electrochemical Model*

The electrochemical model determines the current density and overpotentials at a given cell voltage. The cell voltage $V^{cell}$ is calculated as the combination of open circuit voltage (OCV) $V^{OCV}$, anode activation overpotential $\eta_A^{act}$, cathode activation overpotential $\eta_C^{act}$, anode concentration overpotential $\eta_A^{conc}$, cathode concentration overpotential $\eta_C^{conc}$, and electrolyte overpotential (i.e., ohmic loss in electrolyte), $\eta_E$, as

$$V^{cell} = V^{OCV} - \eta_A^{act} - \eta_C^{act} - \eta_A^{conc} - \eta_C^{conc} - \eta_E \tag{9}$$

$$V^{OCV} = -\frac{\Delta G^0}{2F} + \frac{R_g T}{2F} ln\left(\frac{x_{H_2}}{x_{H_2O}}\left(x_{O_2}\right)^{\frac{1}{2}}\right) \tag{10}$$

To calculate the activation overpotential $\eta^{act}$, the Butler-Volmer expression can be used to describe $\eta^{act} - i$ relationships

$$i = i_0\left[exp\left(\frac{\alpha_a F \eta^{act}}{R_g T}\right) - exp\left(-\frac{\alpha_c F \eta^{act}}{R_g T}\right)\right] \tag{11}$$

where $i_0$ is the exchange current density, $\alpha_a$ and $\alpha_c$ are the anodic and cathodic charge transfer coefficient, respectively. To determine the concentration overpotential $\eta_C$, the following equations can be used:

$$\eta_A^{conc} = -\frac{R_g T}{2F} ln\left(\frac{P_{H_2}}{P_{H_2}^0}\frac{P_{H_2O}^0}{P_{H_2O}}\right) \tag{12}$$

$$\eta_C^{conc} = -\frac{R_g T}{4F} ln\left(\frac{P_{O_2}}{P_{O_2}^0}\right) \tag{13}$$

where $P_i^0$ is an inlet pressure of species $i$. However, concentration overpotential can be neglected if the porous media transport and electrochemical reactions are modeled in detail (Janardhanan and Deutschmann 2006). The electrolyte overpotential $\eta_E$ is dependent on the current density, electrolyte thickness, and temperature which are obtained by

$$\eta_E = \frac{il_E}{\sigma_E} \tag{14}$$

where $l_E$ is electrolyte thickness, $\sigma_E$ is electrolyte conductivity.

## 3 Numerical Implementation

OpenFOAM® solves a set of governing equations using a finite volume scheme written in programming language C++. The main components are: (1) Second order finite volume discretisation on three-dimensional meshes, with polyhedral cell support and capability of handling local mesh refinement, (2) Versatile and transparent representation of partial differential equations in the code, achieved through equation mimicking, (3) A complete set of implicit discretisation operators, and associated matrix and linear solver classes, (4) Massive parallelisation in domain decomposition mode, (5) The capability of dealing with multiple overlapping regions within a single solver, with efficient data transfer between them. Each region can support its own set of fields, e.g., pressure, velocity, species concentration, energy, with associated boundary conditions, (6) Access to libraries of real material properties for single species and their mixtures, (7) Efficient and validated set of heat transfer, fluid flow, chemical reaction, and electrochemistry solvers.

The pressure-momentum coupling is solved using the Pressure Implicit with Splitting of Operators (PISO) algorithm. Linear algebraic systems are solved using Preconditioned bi-conjugate gradient (PBiCG) solver with Diagonal-based Incomplete LU (DILU) preconditioner for velocity, mass fraction, and temperature, and Preconditioned conjugate gradient (PCG) solver with Diagonal-based Incomplete Cholesky (DIC) preconditioner for pressure. Volumetric equations are discretized on the fixed and structured meshes composed of hexahedral element. Three-dimensional (3D) mesh is generated using automatic meshing technique. The meshes are composed of global-region and sub-regions which are split up into individual components. The continuity, momentum and species transport equations are solved on the individual meshes, while energy equation is solved on the global mesh. The calculated properties and variables are passed back and forth between global and individual meshes by mapping function. The physical properties of gases are calculated by accessing to OpenFOAM® library or class library. The solution algorithm proceeds as follows:

1. Create the cell geometry, specify boundary conditions, assign physical properties, and prescribe average current density.
2. Calculate continuity and momentum equation.
3. Calculate Nernst potential, overpotentials, and cell voltage.
4. Calculate sources of mass, species, and energy.
5. Calculate conservation equations of species transport and energy.
6. Repeat steps (2)–(5) until convergence is obtained.

The algorithm is sequential and iterative. The cell voltage is computed by prescribing a designed current density $i^{des}$ as

$$V^{cell} += \alpha \times (\bar{i} - i^{des}) \tag{15}$$

where $\alpha$ is a relaxation factor. The calculations are repeated until the difference between the computed mean current density $\bar{i}$ and designed current density $i^{des}$ is within the convergence criteria.

## 4 Geometry, Mesh and Boundary Condition

For an application, a planar-type anode-supported SOFC is modeled as shown in Fig. 1. The basic unit is composed of interconnects, gas channels, and PEN. The flow channels are embedded within the rectangular cross-section of interconnects. The fuel and air channels are separated by PEN which is composed of five layers: (1) anode substrate layer (ASL), (2) anode functional layer (AFL), (3) electrolyte, (4) cathode functional layer (CFL), and (5) cathode current collector layer (CCCL). Large-scale anode-supported SOFC is modeled that has reaction area of 100 $cm^2$ (10 cm × 10 cm) and channel size of 1 mm × 1 mm with same rib-spacing width as channel width. Co-flow configuration is considered. Through the grid independency test, total mesh of 2.08 million is used for the simulation (Fig. 2). The finer mesh increase the accuracy, but incurs high calculation cost. Neumann boundary condition is applied on the wall boundaries (adiabatic), and Dirichlet boundary condition is imposed on the inlet, top and bottom boundaries (constant temperature).

The cell geometry can be constructed using commercial software or can be virtually generated using numerical algorithm. Here, we create a geometry and mesh the entire domain using the mesh generation utility *blockMesh* and the geometry generation utility *cellSets*. A directory file named *blockMeshDict* located in the run/constant/polymesh directory. The *blockMesh* reads this dictionary, and writes out the mesh data to *points* and *faces*, *cells* and *boundary* files in the same directory.

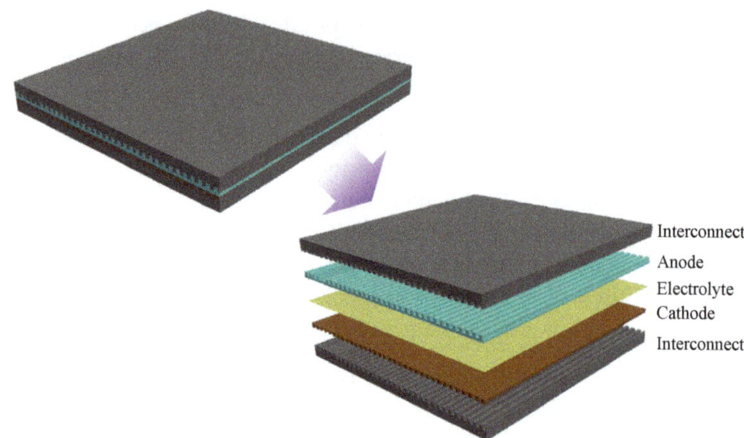


**Fig. 1** Schematic diagram of planar-type anode-supported SOFC. Computational domain includes interconnects, channels, anode, electrolyte and cathode. Reproduced from Jeon (2019) with permission from Elsevier

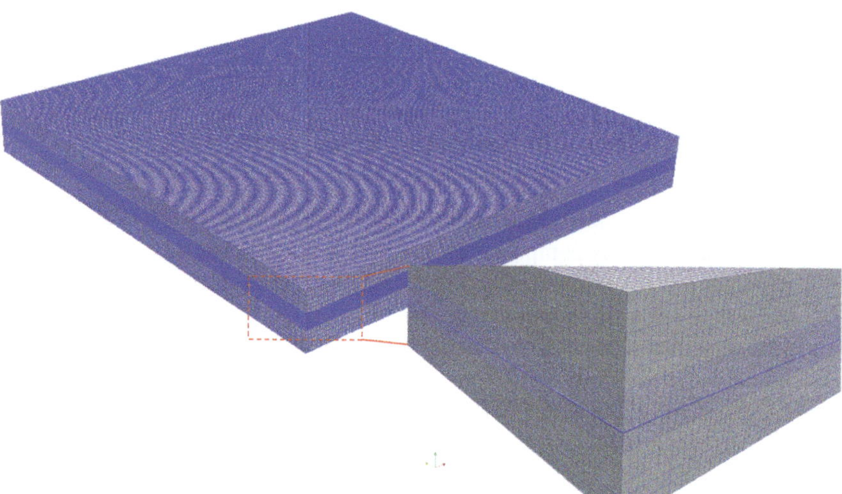

**Fig. 2** Tessellated meshes for a planar-type anode-supported SOFC. The zoom-in view illustrates the structured-rectangular meshes

The *cellSets* makes up the cells of each region, and generate the meshes for each region. The *cellSets* for the regions are specified in /run/config directory.

## 5 Results and Discussion

This section discusses the computed results of anode-supported SOFCs. Geometry details, cell properties and operating condition that used in simulation are listed in Table 1. The user can visualize the results using the post-processing tool, such as *ParaView* and *EnSight*.

In this simulation, we employed the most common choice materials for SOFC components, which are yittria stabilized zirconia (YSZ) for electrolyte, Ni–YSZ cermet for anode, and lanthanum strontium manganite (LSM)–YSZ cermet for cathode. To calculate the activation overpotential $\eta^{act}$, the modified Butler-Volmer expression which is taken from Janardhanan and Deutschmann (2006) and Zhu et al. (2005) is used as.

$$i_A = i_{0,A}\left[exp\left(\frac{(\beta_a + 1)F\eta_A^{act}}{R_g T}\right) - exp\left(\frac{\beta_c F\eta_A^{act}}{R_g T}\right)\right] \tag{16}$$

$$i_{0,A} = i_{H2}^{*}\left[\frac{\left(P_{H2}/P_{H2}^{*}\right)^{1/4}(P_{H2O})^{3/4}}{1 + \left(P_{H2}/P_{H2}^{*}\right)}\right] \tag{17}$$

$$i_{H2}^{*} = K_{H2}exp\left(-\frac{E_{H2}}{RT}\right) \tag{18}$$

$$P_{H2}^{*} = \frac{A_{des}\Gamma^{2}\sqrt{2\pi RTM_{H2}}}{\gamma_0}exp\left(-\frac{E_{des}}{R_g T}\right) \tag{19}$$

where $\beta_c = 0.5$, $\beta_a + \beta_c = 1$, $K_{H2} = 2.07 \times 10^5 Am^{-2}$ and $E_{H2} = 87.8KJmol^{-1}$. The pre-exponent factor $A_{des} = 5.59 \times 10^9 scm^2mol^{-1}$, the surface site density $\Gamma = 2.6 \times 10^{-9} molcm^{-2}$, the activation energy $E_{des} = 88.12KJmol^{-1}$, and the

**Table 1** Properties and operating conditions

| Properties | | | | | | |
|---|---|---|---|---|---|---|
| | $l$ (μm) | $d_p$ (μm) | $\varepsilon$ | $\tau$ | $\kappa(W/m\,k)$ | $C_p$ (J/kg·K) |
| ASL | 1000 | 2 | 0.5 | 3 | 3 | 1000 |
| AFL | 20 | 1 | 0.25 | | | |
| Electrolyte | 10 | – | – | – | 2.4 | 550 |
| CLF | 20 | 1 | 0.25 | 3 | 3 | 750 |
| CCCL | 50 | 6 | 0.5 | | | |
| Interconnect | 5000 | – | – | – | 24 | 660 |
| Operating conditions | | | | | | |
| $x_{H_2}^{o}, x_{O_2}^{o}$ | 0.97, 0.21 | | | | | |
| $u_{fuel}, u_{air}$ | 1 m/s, 2 m/s | | | | | |

sticking probability $\gamma_0 = 0.01$.

$$i_C = i_{0,C}\left[exp\left(\frac{\beta_a F \eta_C^{act}}{R_g T}\right) - exp\left(\frac{\beta_C F \eta_C^{act}}{R_g T}\right)\right] \quad (20)$$

$$i_{0,C} = i_{O2}^{*}\frac{\left(P_{O2}/P_{O2}^{*}\right)^{1/4}}{1 + \left(P_{O2}/P_{O2}^{*}\right)^{1/2}} \quad (21)$$

$$i_{O2}^{*} = K_{O2}exp\left(-\frac{E_{O2}}{R_g T}\right) \quad (22)$$

$$P_{O2}^{*} = A_{O2}exp\left(-\frac{E_{O2}}{R_g T}\right) \quad (23)$$

where $K_{O2} = 5.19 \times 10^4 Am^{-2}$, $E_{O2} = 88.6KJmol^{-1}$, $A_{O2} = 4.9 \times 10^8 atm$ and $E_{O2} = 200KJmol^{-1}$. Note that Butler-Volmer expression established by Janardhanan and Deutschmann (2006) was modified at the temperature range of 600 – 800 °C. The electrolyte conductivity of YSZ which are obtained by Ferguson et al. (1996) is

$$\sigma_E = 3.34 \times 10^4 exp\left(-\frac{10300}{T}\right) \quad (24)$$

Figure 3 shows the simulated performance with contribution of component overpotentials at 700 °C. The performance decreases as the current density increases. The contributions of anode overpotential and cathode overpotential are large at low current density, but the contribution of cathode overpotential becomes dominant at high current density. The contribution of electrolyte overpotential is not significant which is generally large during the SOFC operation. This is mainly attributed to thin electrolyte layer.

The simulated result also provides the detailed information for dependent variables. Figure 4 shows illustrative examples of the distributions of current density, temperature, species mass fraction and component overpotential at 700 °C. As shown, current density is high at the inlet where the mole fraction of reactants is high, and becomes lower along the electrode length. Current density distribution affects temperature distribution which is high at the inlet and low at the outlet. The mole fractions of hydrogen and oxygen are high at the inlet, and becomes lower along the electrode length. Hence, activation overpotentials are high at the inlet, and concentration overpotentials are high at the outlet. Interestingly, several straight lines are observed all the figures likely due to non-uniform electrochemical reaction. This is ascribed to non-uniform oxygen distribution, in which oxygen mole fraction is observed high at the channel area, but low at the rib area. These severely non-uniform distributions are attributed to the thin cathode layer. To avoid non-uniform distributions,

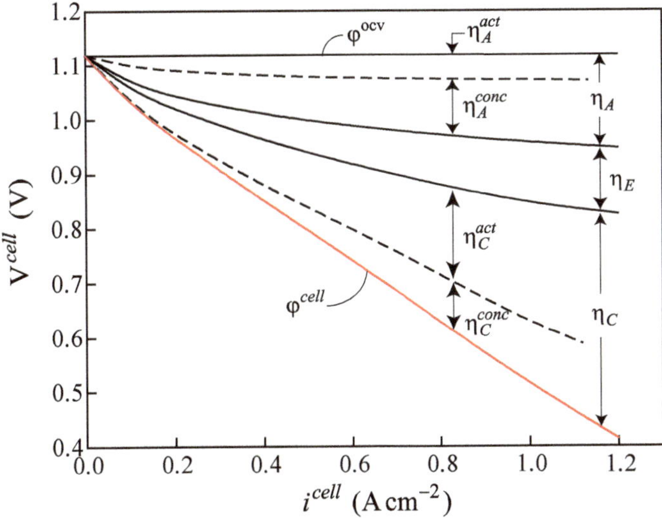


**Fig. 3** Simulated performance and contribution of component overpotentials to the overall potential loss of anode-supported SOFC at 700 °C. Reproduced from Jeon (2019) with permission from Elsevier

blowing the air with high stoichiometric flow rate or optimum design of cathode microstructure is desirable (Fig. 5).

**Oxygen depletion during calculation**

When operating anode-supported SOFC, the oxygen can be depleted at high current densities. Although the actual SOFC can operate regardless depletion, the model calculation will fail to converge. Once the oxidant is starved, the mole fraction of oxygen gets zero or negative value that affects the calculation of concentration overpotential in Equation 26, resulting in mathematical error. This typically occurs under the rib where the oxidant is difficult to reach due to thin cathode layer, as seen in Figure 5. To avoid this type of calculation failure, zero or negative mole fractions must be replaced to small positive number. Slow converging with low relaxation factor will also be helpful.

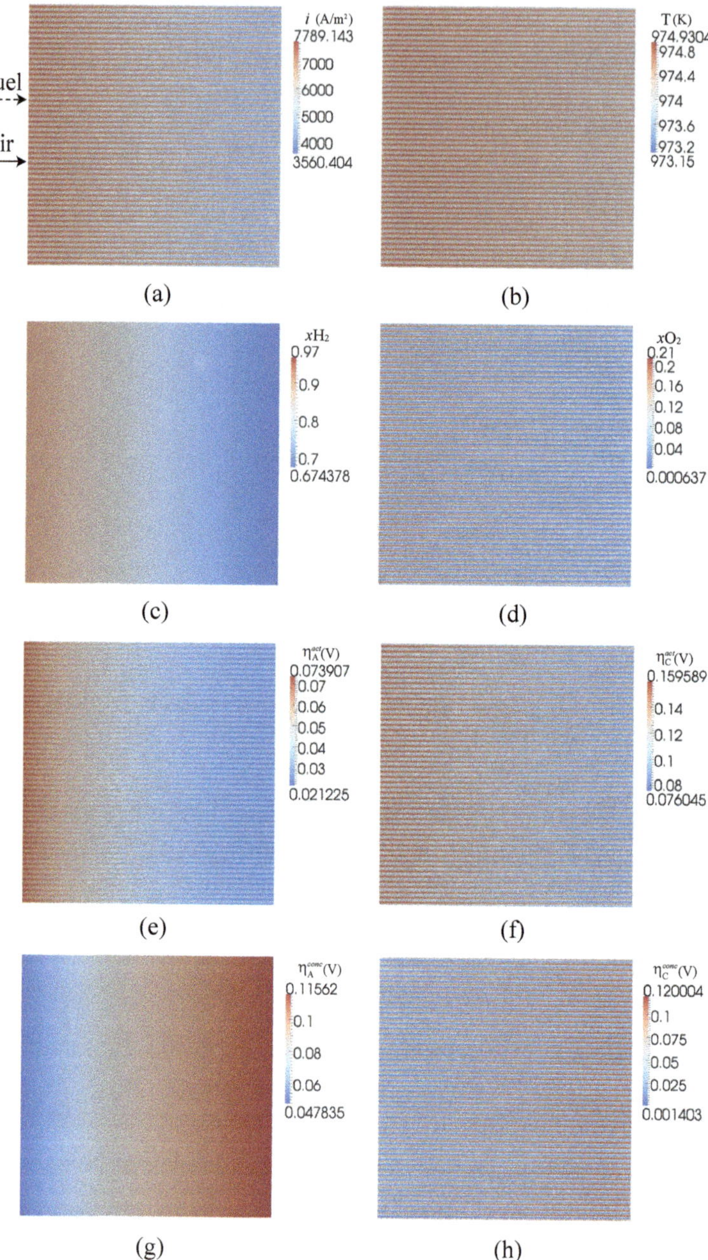


**Fig. 4** Distributions of **a** current density, **b** temperature, **c** hydrogen mole fraction, **d** oxygen mole fraction, **e** anode activation overpotential, **f** cathode activation overpotential, **g** anode concentration overpotential and **h** cathode concentration overpotential on electrode surfaces at 0.6 A/cm$^2$ and 700 °C. Reproduced from Jeon (2019) with permission from Elsevier

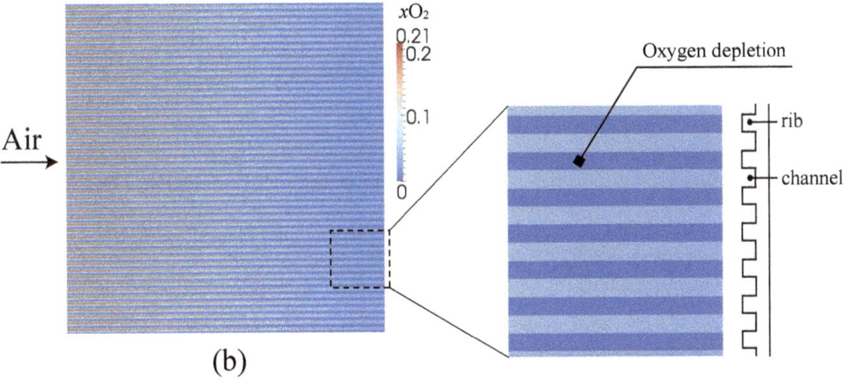


**Fig. 5** Oxygen mole fraction distributions on CFL surface at 1.0 A/cm$^2$ and 700 °C. Reproduced from Jeon (2019) with permission from Elsevier

**Acknowledgements** This work was supported by the Dongguk University Research Fund of 2020.

## References

Achenbach E (1994) Three-dimensional and time-dependent simulation of a planar solid oxide fuel cell stack. J Power Sour 49:333–348

Beale SB, Le AD, Roth HK, Pharoah JG, Choi H-W, de Haart LGJ, Froning D (2011) Numerical and experimental analysis of a solid oxide fuel cell stack. ECS Trans 35(1):935–943

Beale S, Choi H-W, Pharoah JG, Roth HK, Jasak H, Jeon DH (2016) Open-source computational model of a solid oxide fuel cell. Comp Phys Commu 200:15–26

Beale SB, Reimer U, Froningm D, Jasak H, Andersson M, Pharoah JG, Lehnert W (2018) Stability issues of fuel cell models in the activation and concentration regimes. J Electrochem Energy Conv Storage 15:041008

Bertei A, Nucci B, Nicolella C (2013) Microstructure modeling for prediction of transport properties and electrochemical performance in SOFC composite electrodes. Chem Eng Sci 101:175–190

Chan SH, Xia ZT (2001) Anode micro model of solid oxide fuel cell. J Electrochem Soc 148:A388–A394

Costamagna P, Costa P, Antonucci V (1998) Micro-modelling of solid oxide fuel cell electrodes. Electrochim Acta 43:375–394

Ferguson JR, Fiard JM, Herbin R (1996) Three-dimensional numerical simulation for various geometries of solid oxide fuel cells. J Power Sources 58:109–122

Hajimolana SA, Hussain MA, Ashri Wan Daud WM, Soroush M, Shamiri A (2011) Mathematical modeling of solid oxide fuel cells: a review. Renew Sust Energy Rev 15:1893–1917

Hussain MM, Li X, Dincer I (2006) Mathematical modeling of planar solid oxide fuel cell. J Power Sour 161:1012–1022

Janardhanan VM, Deutschmann O (2006) CFD analysis of a solid oxide fuel cell with internal reforming: Coupled interactions of transport, heterogeneous catalysis and electrochemical processes. J Power Sources 162:1192–1202

Jeon DH (2019) Computational fluid dynamics simulation of anode-supported solid oxide fuel cells with implementing complete overpotential model. Energy 188:116050

Kone J-P, Zhang X, Yan Y, Hu G, Ahmadi G (2018) CFD modeling and simulation of PEM fuel cell using OpenFOAM. Energy Procedia 145:64–69

Kong W, Gao X, Liu S, Su S, Chen D (2014) Optimization of the interconnect ribs for a cathode-supported solid oxide fuel cell. Energies 7:295–313

Le AD, Beale SB, Pharoah JG (2015) Validation of a solid oxide fuel cell model on the international energy agency benchmark case with hydrogen fuel. Fuel Cells 15:27–41

Nagata S, Momma A, Kato T, Kasuga Y (2001) Numerical analysis of output characteristics of tubular SOFC with internal reformer. J Power Sour 101:60–71

Nam JH, Jeon DH (2006) A comprehensive micro-scale model for transport and reaction in intermediate temperature solid oxide fuel cell. Electrochim Acta 51:3446–3460

Ni M (2012) Modeling of SOFC running on partially pre-reformed gas mixture. Intl J Hydrog Energy 37:1731–1745

Nishida RT, Beale SB, Pharoah JG, de Haart LGJ, Blum L (2018) Three-dimensional modelling and experimental validation of the Julich Mark-F solid oxide fuel cells. J Power Sour 373:203–210

OpenFOAM®: the open source CFD toolbox available from http://www.openfoam.com

Pramuanjaroenkij A, Kakac S, Zhou XY (2008) Mathematical analysis of planar solid oxide fuel cells. Intl J Hydrog Energy 33:2547–2565

Recknagle KP, Williford RE, Chick LA, Rector DR, Khaleel MA (2003) Three-dimensional thermos-fluid electrochemical modeling of planar SOFC stacks. J Power Sour 113:109–114

Vijay P, Tade MO, Shao Z, Ni M (2017) Modelling the triple phase boundary length in infiltrated SOFC electrodes. Intl J Hydrog Energy 42:28836–28851

Yakabe H, Ogiwara T, Hishinuma M, Yasuda I (2001) 3-D model calculation for planar SOFC. J Power Sour 102:144–154

Zhang S, Reimer U, Beale SB, Lehnert W, Stolten D (2018) Modeling polymer electrolyte fuel cells: a high precision analysis. Appl Energy 233–234:1094–1103

Zhu H, Kee RJ, Janardhanan VM, Deutschmann O, Goodwin DG (2005) Modeling elementary heterogeneous chemistry and electrochemistry in solid-oxide fuel cells. J Electrochem Soc 152:A2427–A2440

# Transport Modeling of High Temperature Fuel Cell Stacks

**Shidong Zhang and Robert Nishida**

**Abstract** Multiple electrochemical cells are arranged into stacks to increase power density. Electrochemical performance of stacks is significantly affected by the design of manifolds that carry reactants to, and reaction products from, each cell in the stack. Flow, pressure, and heat mal-distributions may occur, which leads to variations in the local current density, overall reductions in stack power output and, in some cases, localized damage or degradation effects which are not captured by single cell or purely hydrodynamic models. The modeling of large-scale stacks, necessary for detailed analyses and design, requires the making of assumptions in transport equations in localized regions to resolve phenomena of flow, species and heat transport within reasonable computation times. In this chapter, a distributed resistance analogy (DRA) is described in which the transport equations are averaged by using a first-order rate term to replace the second-order diffusion term in the Navier–Stokes, heat transfer, and species transport equations, using a friction factor, Nusselt number heat transfer resistance, and Sherwood number species transport resistance, respectively. The technique reduces the computational mesh required and so also the required computation time by up to two orders of magnitude. The implementation of the DRA techniques in OpenFOAM are described in detail for modelling real solid oxide fuel cell stacks and high-temperature polymer electrolyte fuel cell stacks. Numerical verification and experimental validation are systematically conducted with good agreement, providing confidence in the accuracy of the model and its implementation in OpenFOAM.

---

The original version of this chapter was revised: The ESM material has been updated. The correction to this chapter is available at https://doi.org/10.1007/978-3-030-92178-1_10

**Supplementary Information** The online version contains supplementary material available at https://doi.org/10.1007/978-3-030-92178-1_9.

---

S. Zhang (✉)
Institute of Energy and Climate Research (IEK), Forschungszentrum Jülich GmbH, 52428 Jülich, Germany
e-mail: s.zhang@fz-juelich.de

R. Nishida
Department of Mechanical Engineering, University of Alberta, Edmonton, AB T6G 2G8, Canada

© Springer Nature Switzerland AG 2022, corrected publication 2022

S. Beale and W. Lehnert (eds.), *Electrochemical Cell Calculations with OpenFOAM*, Lecture Notes in Energy 42, https://doi.org/10.1007/978-3-030-92178-1_9

## 1 Introduction

Multiple electrochemical cells are layered into stacks composed of dozens or even hundreds of single cells (Lüke et al. 2012; Janßen et al. 2013, 2018, 2017) in order to increase power density. Manifolds are designed to distribute reactant species to, and remove excess heat and reaction products from, each individual electrochemical cell. Within each cell, depending on the geometry of the manifolds, local current density tends to vary significantly (Yuan 2008; Costamagna et al. 1994). Localized areas that are starved of fuel become inactive, cooler and have lower localized reaction rates (Recknagle et al. 2003), which can strongly limit the stack's overall performance and can, in some cases, cause damage to it (Costamagna et al. 1994; Kulikovsky 2019). The distribution of temperature within the stack has a strong effect on electrochemical performance, electronic and ionic conduction, and transport properties, which are especially relevant for high-temperature fuel cells (Kulikovsky 2019; Achenbach 1994; Nishida et al. 2016). Therefore, it is vital that the inter-dependent phenomena of flow, species transport and heat transfer are carefully managed to maximize the electrochemical performance of each cell and the overall stack.

Figure 1 shows a five-cell, high-temperature polymer electrolyte fuel cell (HT-PEFC) stack that has operating temperatures of 140–180 °C and an 18-cell solid oxide fuel cell (SOFC) stack that has operating temperatures of 600–800 °C. Polymer electrolyte fuel cells (PEFCs) convert the chemical potential of hydrogen and oxygen directly into electricity with high efficiency and low pollution (Weber et al. 2014; Andersson et al. 2016; Zhang and Jiao 2018). Due to their higher operating temperature, HT-PEFCs have a higher tolerance for carbon monoxide (CO) (Li et al. 2009; Liu et al. 2016) than conventional PEFCs and are able to operate using low-purity gases, e.g., reformate (Steinberger-Wilckens and Lehnert 2010). A phosphoric acid-doped polybenzimidazole membrane is often employed in the HT-PEFCs, and therefore gas humidification is not necessary to maintain high protonic conductivity and the humidification system is avoided (Li et al. 2009). The high-temperature operation of SOFCs enables the internal reforming of fuels other than pure H2, such as methane, propane, butane, and other light hydrocarbons. Moreover, the kinetic activity and charge transport are both improved at high temperatures. As a result, SOFCs have a relatively high-power density and commercially-available systems can reach upwards of 60% electrical efficiency. Furthermore, SOFCs produce high-quality waste heat in their exhaust stream that can be used in combined heat and power applications to reach very high system efficiencies of greater than 80% (Hirschenhofer et al. 1994).

Broadly speaking, each of a stack's electrochemical cells consist of current-conducting material through which fuel and air flow on either side of an electrolyte assembly. The gases flow through either straight or serpentine channels that are machined into plates, or through a wire mesh to distribute reactants to the electrolyte assembly in each cell and remove heat and reaction products. Within each electrolyte assembly, layers of porous materials distribute gas to and from the reaction sites at which the electro-chemical reactions occur. Importantly, in fuel cell stacks, manifolds transport fuel and air to, and reaction products from, each individual cell. Flow,

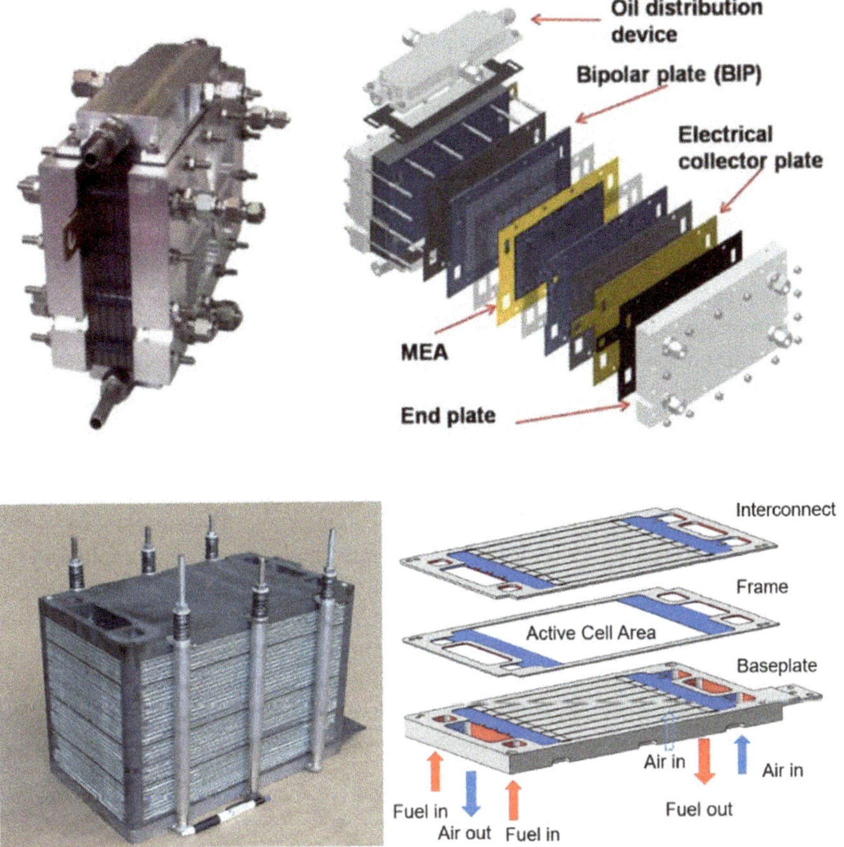


**Fig. 1** Top: Five-cell polymer electrolyte fuel cell stack (200 cm$^2$ active area per cell; Forschungszentrum Jülich), from Kvesic et al. (2012) with permission; and bottom: 18-cell solid oxide fuel cell stack (Mark-F; 400 cm$^2$ active area per cell; Forschungszentrum Jülich), from (Nishida et al. 2018) with permission

species, and heat transport from manifolds at the meter scale, to cell channels at the millimeter scale, and porous transport layers at the micro-scale, represent a challenging multi-scale transport problem that is intimately connected to electrochemical reactions.

In order to evaluate the performance of systems or components of fuel cell stacks, experiments are conducted by varying such operating conditions as temperature, pressure, and the stoichiometric factor, amongst many others (Ubong et al. 2009; Liu 2019). To complement and help guide these experiments, mathematical models of stacks have been developed, ranging from zero-dimensional (0-D) system level models (Arsalis et al. 2011a, b) to detailed ones (Sun et al. 2018; Liu et al. 2006), which consider a range of physico-chemical phenomena in three-dimensional (3-D) simulations. Due to computational limitations, analyses of fuel cell stacks commonly consider either fluid transport or electrochemical phenomena only, neglecting any interdependence. Models that consider fluid flow, heat transfer and electrochemistry are generally limited to single cells. The modeling of large-scale stacks requires assumptions to be made regarding the transport equations in localized regions to enable the models to capture flow, species and heat transport in reasonable computation times. In order to strike a balance between accuracy and computational effort, methods for volume-averaging (Nishida et al. 2016; Roos et al. 2003; Beale and Zhubrin 2005; Kvesic et al. 2012; Nandjou et al. 2016a, b) are often applied in stack level simulations.

In this chapter, the distributed resistance analogy (DRA) is described in which volume averaging is applied to the transport equations, thereby reducing the computational requirements (Beale and Zhubrin 2005). In the flow channels, porous media and the reaction regions of the electrochemical stack, the momentum, heat, and species transport equations are averaged by using a first-order rate term to replace the second-order diffusion term. The technique reduces the number of computational cells and thus the required computation time by up to two orders of magnitude. The DRA model, which was originally developed by Patankar and Spalding (PATANKAR, S.V. and D.B. Spalding 1972) for heat exchanger analysis, was first applied to fuel cells by Beale and Zhubrin (Beale and Zhubrin 2005). In this chapter, the DRA techniques described were executed in OpenFOAM® (https://openfoam.org/) for both theoretical and models of real solid oxide fuel cell stacks (Nishida et al. 2016) and HT-PEFC stacks (Sun et al. 2018) for experimental validation, and to demonstrate stack-scale effects not captured by single-cell or purely hydrodynamic models. For model verification, the detailed numerical method (DNM) is also applied in which the full transport equations are solved (Beale et al. 2016; Zhang et al. 2019).

## 2 Numerical Methods

A detailed numerical method (DNM) considers macroscopic geometric details in modeling and simulations. A set of conformal meshes are generated with body-fitting applied to different parts of the fuel cell stacks such as the cell channels and distinct layers of electrochemical cells. This approach requires a computational mesh with high resolution, particularly near the geometrical features. Therefore, intensive computational effort is unavoidable as discussed in Zhang et al. (Zhang et al. 2019). The distributed resistance analogy (DRA) enables mesh resolution and computation time to be reduced, yet accurately capture the transport phenomena relevant to large-scale stacks. This section outlines the details of the DRA model with reference to the DNM.

## 2.1 Governing Equations of the DRA

Fuel cell operation involves inter-dependent phenomena of heat and mass transfer, momentum transport, species redistribution, and electrochemical reactions. The steady-state form of the governing transport equations can be expressed as follows for the transport of a generic scalar, $\varphi$, of mass, heat, or species:

$$\nabla \cdot (\rho \boldsymbol{u} \boldsymbol{\varphi}) = \nabla \cdot (\Gamma \nabla \boldsymbol{\varphi}) + \sum_{\mathrm{j}} \alpha_{\mathrm{j}} (\boldsymbol{\varphi} - \boldsymbol{\varphi}_{\mathrm{j}}) + S \tag{1}$$

where, the four terms shown are (from left to right), convection, diffusion, interphase-transfer, and source terms, respectively; $\rho$ denotes fluid/solid density, $u$ represents the velocity, $\Gamma$ refers to the viscosity/thermal conductivity/diffusion coefficient, and $S$ is the source/sink term; the interphase transfer term, $\sum_j \alpha_j (\varphi - \varphi_j)$, is the rate term of the DRA which supplants the diffusion term in select regions of the stack as described in detail below. The DNM applies the diffusion term throughout the geometry and does not use the rate term of Eq. (1). Each transport equation is converted into a linear algebraic formulation and solved using the finite volume method (FVM)-based computational fluid dynamics (CFD) package, OpenFOAM (https://openfoam.org/). The flow is considered laminar and incompressible, contained within the manifolds and cell channels and the electrolyte is considered impermeable to any gases. For heat transfer, thermal radiation effects are not considered internal to the stack, local heat sources are applied as volumetric source terms in the electrolyte assembly rather than at localized microscopic electrochemical reaction sites, and heating in the interconnects or porous transport (gas diffusion) layers is considered negligible. For conservation of species, Fick's law is applied. For momentum, heat and species transport, the steady-state form of each equation is applied and coupled with equations for localized electrochemical reactions at each electrochemical cell.

Electrochemical cells are layered into stacks of repeat unit cells. A PEN (Positive-electrode, Electrolyte, Negative-electrode) for SOFCs or MEA (membrane electrode assembly) for HT-PEFCs is compressed between interconnects or bipolar plates which contain gas distribution channels. A cross-section of a single electrochemical cell of an SOFC is shown in Fig. 2 in which the current collector layer (or gas diffusion layer), function layer (or catalyst layer), and electrolyte (or membrane) make up the PEN (or membrane assembly). In the DRA, a volume averaging method is applied; therefore, the detailed geometry of each channel or layer of membrane assembly is not resolved. The detail of a unit cell is averaged within a computational domain while each of four regions of the domain (i.e. interconnect, membrane assembly, air and fuel) is assigned a volume fraction. Equation (1) is solved for each of the four regions with inter-phase transfer terms representing heat or species transport between each region. For example, convective heat transfer takes place between the gases and the porous media and interconnects and gases diffuse in/out of the porous media to/from the functional layer where electrochemical reactions occur.

From Eq. (1), the mass continuity equation is applied in the following form:

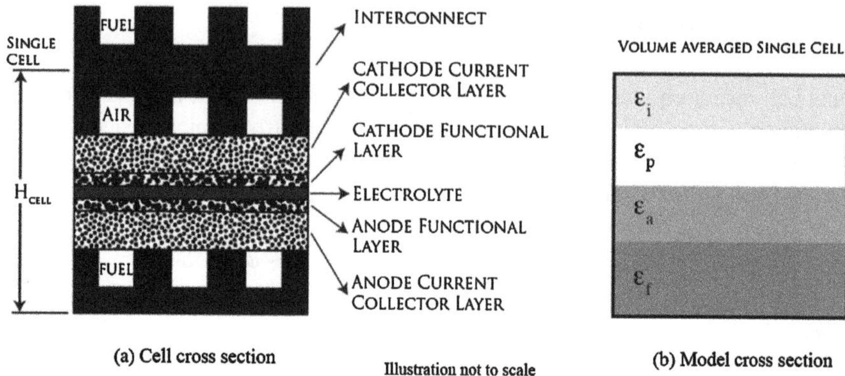


**Fig. 2** Cross-section of: **a** detailed components of a solid oxide fuel cell; and **b** a simplified computational domain for applying the distributed resistance analogy

$$\nabla \cdot (\varepsilon \rho \boldsymbol{u}) = \boldsymbol{\varepsilon} \dot{m} \tag{2}$$

where $\varepsilon$ is the porosity of the fluid region, $\rho$ means the mixture density, $\boldsymbol{u}$ represents the interstitial velocity, and $\dot{m}$ is the mass source/sink. The pressure and velocity distributions may be calculated using the momentum transport equation as follows:

$$\nabla \cdot \left(\varepsilon \rho \frac{\boldsymbol{u}\boldsymbol{u}}{\tau}\right) = -\frac{\varepsilon}{\tau^2} \nabla p + \nabla \cdot \left(\varepsilon \frac{\mu}{\tau^2} \nabla \boldsymbol{u}\right) - \frac{\varepsilon}{\tau^2} \boldsymbol{F}_D \tag{3}$$

where $p$ is the pressure, $\mu$ represents the kinematic viscosity, $\tau$ means the tortuosity ($\tau = 1$ in straight channels and $\tau > 1$ in serpentine channels), and $\boldsymbol{F}_D$ is the drag term due to wall friction, which can be computed as:

$$\boldsymbol{F}_D = \xi \tau \frac{2}{D_h^2} \mathrm{fRe}_h \mu \boldsymbol{u} \tag{4}$$

where $D_h$ is the hydraulic diameter, f is the Fanning friction factor, $\mathrm{Re}_h$ represents the Reynolds number, and $\xi$ is a correction factor. The values of $\mathrm{fRe}_h$ can be obtained from the work of Shah and London (Shah and London 1978). The values of $\xi$ vary with the design of the flow path, e.g., $\xi = 1.0$ in the straight channels and $\xi > 1.0$ in the serpentine ones. The species transport equation is:

$$\nabla \cdot (\varepsilon \rho \boldsymbol{u} y_{\mathrm{i}}) = \nabla \cdot \left(\varepsilon \frac{\Gamma_{\mathrm{i}}}{\tau^2} \nabla y_{\mathrm{i}}\right) + \varepsilon \dot{m}_{\mathrm{i}}'' \tag{5}$$

where $y_{\mathrm{i}}$ is the mass fraction of species i, $\Gamma$ represents the binary diffusion coefficient, and $\dot{m}_{\mathrm{i}}''$ is the mass source/sink of species i. The enthalpy equation yields:

$$\nabla \cdot (\varepsilon \rho \boldsymbol{u} T) = \nabla \cdot \left(\varepsilon \frac{k}{\tau^2} \nabla T\right) + \sum\nolimits_{\mathrm{j}} \alpha_{\mathrm{j}} (T - T_{\mathrm{j}}) + \varepsilon \dot{q}''' \tag{6}$$

where $T$ is the temperature, $k$ is the thermal conductivity, $\dot{q}'''$ is the heat source/sink that applies only in the region of the membrane assembly, and $\alpha_j$ represents the heat transfer coefficient between the region in which the equation is being solved and a neighboring region $j$. The calculation of $\alpha_j$ yields:

$$\alpha V = UA \tag{7}$$

where $V$ is the volume of the region, $A$ represents the contact area between two adjacent regions, and $U$ denotes the surface heat transfer coefficient, which is typically given by:

$$U = \frac{1}{h} + \left(\frac{H}{Sk_s}\right) \tag{8}$$

where $S$ is the conduction shape factor, $H$ denotes the thickness of the solid regions, and $h$ represents the heat transfer coefficient for the fluid regions:

$$h = \frac{\mathrm{Nu}k_f}{D_\mathrm{h}} \tag{9}$$

where Nu is the Nusselt number (Beale 2015). The source/sink terms in the governing equations are listed in Table 1.

To describe the electrochemical reactions in fuel cell stacks, a Kirchhoff-Ohm relation is applied:

$$V_\mathrm{cell} = E - \eta - i''R \tag{10}$$

**Table 1** Source and sink terms in the governing equations

| Description | Symbol | Value |
|---|---|---|
| Species mass source/sink | $\dot{m}_i''$ | $O_2$: $-\frac{M_{O_2} i''}{4F H_{air}}$ <br> $H_2$: $-\frac{M_{H_2} i''}{2F H_\mathrm{fuel}}$ <br> $H_2O$: $\frac{M_{H_2O} i''}{2F H_\mathrm{air}}$ |
| Total mass source/sink | $\dot{m}''$ | Air: $\begin{pmatrix} \dot{m}''_{H_2O} + \dot{m}''_{O_2} & SOFC \\ \dot{m}''_{O_2} & PEFC \end{pmatrix}$ <br> Fuel: $\begin{pmatrix} \dot{m}''_{H_2} & SOFC \\ \dot{m}''_{H_2O} + \dot{m}''_{H_2} & PEFC \end{pmatrix}$ |
| Heat source | $\dot{q}'''$ | $\frac{i''\left(\frac{\Delta H}{2F} + V_\mathrm{cell}\right)}{H_\mathrm{MEA}}$ |

where $V_{cell}$ is the cell output voltage, $\eta$ represents the activation overpotential that is negligible in SOFCs and a factor in HT-PEFCs, $i''$ denotes the local current density, $R$ represents the area-specific resistance, and $E$ refers to the Nernst potential:

$$E = E^0 + \frac{R_g T}{2F} \ln\left(\frac{X_{H_2} X_{O_2}^{0.5}}{X_{H_2O}}\right) \tag{11}$$

where $E^0$ means the reference potential, $R_g$ is the universal gas constant, $F$ is Faraday's constant, and $X$ represents the molar fraction of each species on the surface of the electrolyte, also referred to as the wall value (Beale 2015). The activation overpotential, $\eta$, if considered, can be expressed via a Tafel relation:

$$i'' = i_0''\left(\frac{X_{O_2}}{X_{ref}}\right) e^{\frac{\alpha_0 F}{R_g T}\eta} \tag{12}$$

where $\alpha_0$ is the transfer coefficient and $i_0$ refers to the exchange current density.

## 2.2 Computational Procedure

All of the numerical discretization and computations were performed within the OpenFOAM platform. The governing equations were discretized into linear algebraic expressions. They formed two types of matrix systems, symmetrical and asymmetrical. The former was usually solved via an incomplete Cholesky preconditioning (ICCG) method, and the latter with a Biconjugate gradient solver Bi-CGStab method. The size of these matrices is the square of the number of elements in a computational mesh. Hence, a highly resolved mesh requires more computational effort than a coarse one for the same solver. In light of this, calculations of the DRA cases were performed on a single personal computer and completed within a few hours, whereas the DNM requires significantly more computations as detailed below.

By comparison, to conduct the DNM simulations of a HT-PEFC stack (Zhang 2019), for example, the high-performance computing facility, JARA-HPC, was used. Each case takes approximately 36 h in parallel with 900 cores. In the five-cell short PEFC stack, air and pure hydrogen were supplied from the cathode and anode inlets, respectively, with stoichiometric factors of 2/2. The stack was cooled with engine oil at a constant flow flux of 1.5 L min$^{-1}$. The operating temperature was maintained at 160 °C, with the same temperatures for the air, hydrogen, and oil inlets. Additional model parameters are displayed in Table 2.

**Table 2** Operating conditions and model parameters

| Description | Symbol | Units | Values for PEFC | Values for SOFC |
|---|---|---|---|---|
| Pressure | p | Pa | Air, Fuel, Oil: 101,325 | Air, Fuel: 101,325 |
| Reference potential | $E^0$ | V | 1.15 | |
| Transfer coefficient | $\alpha$ | | 0.7 | – |
| Exchange current density | $i_0''$ | $\mathrm{A\,m^{-2}}$ | 0.5 | – |
| Area specific resistance | 0.1 | $\Omega\,\mathrm{cm}^2$ | 0.1 | Polynomial (Jeon et al. 2010) |
| Tortuosity | $\tau$ | | Straight: 1.0<br>Serpentine: 9.02 | 1.0 |
| Volume fraction | $\varepsilon$ | | Air: 0.075<br>Fuel: 0.075<br>BPP: 0.67<br>Oil: 0.08<br>MEA: 0.0075 | Air: 0.0665<br>Fuel: 0.0665<br>PEN: 0.115<br>Interconnect: 0.752 |
| Shape factor | $S$ | | Straight: 0.42<br>Serpentine: 0.51 | Anode: 0.65<br>Cathode: 0.2 |
| Sherwood number | Sh | | 2.78 | 1.65 |
| Nusselt number | Nu | | 3.5 | 3.09 |

## 2.3 *Model Implementation of Distributed Resistance Analogy in OpenFOAM*

The momentum, species, and heat transport equations are commonly solved in OpenFOAM using standard solvers. Additional complexity is introduced by customizing transport through porous media, coupling transport equations through variable transport properties and by capturing interactions between solid and fluid regions. However, the distributed resistance analogy requires inter-phase transfer terms to provide direct coupling of cell-center values between different regions or phases (i.e. fuel, air, interconnect, electrolyte and sometimes oil), a method which is not commonly encountered. Therefore, the following section focusses on the customized implementation of the DRA in OpenFOAM.

This work employs a multiply shared space method (MUSES) to enable the coupling of inter-phase transfer terms between different sub-regions in the computational domain in OpenFOAM (Beale and Zhubrin 2005). With the DRA, geometric details of the repeating unit cell are partially neglected, which results in a pseudo-3D assembly for each cell, as shown by Fig. 3. Some parts of the physical spaces are shared by different subregions, e.g. the green area represents the active area (electrolyte), flow fields of air and fuel, and the ribs of the interconnect. The heat and mass transfer between these regions are also exhibited. It can be seen that electrolyte region is the most critical part that communicates between air, fuel, and interconnect. Mass transfer occurs between electrolyte and air/fuel resulting from the electrochemical

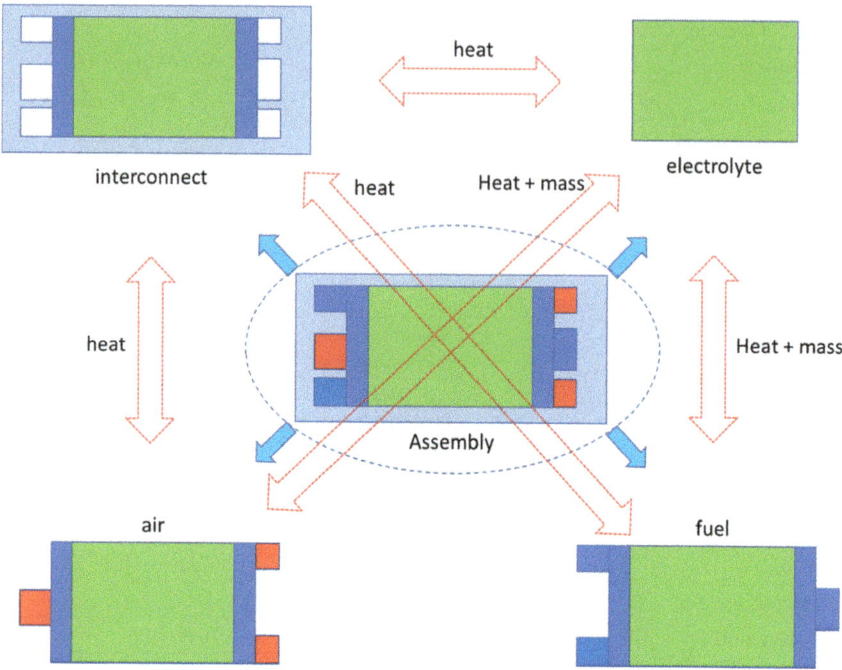


**Fig. 3** Schematic of different regions and the heat and mass transfer between them

reactions via species diffusion from the channel (bulk value) to the ultra-thin catalyst electrodes (wall value), or vice versa. There exists no direct connection between air and fuel.

In OpenFOAM, it is common to use 'parent' and 'child' meshes to solve transport equations in different regions of a geometry depending on requirements. For example, a global temperature profile could be solved on a parent mesh which encompasses transport properties of both solid and fluid regions, whereas the velocity profile generated by Navier–Stokes equations and used in the convective heat transfer equation may only be solved in a child mesh in fluid regions.

The MUSES technique developed by Beale et al. (Beale et al. 2016) was adapted to OpenFOAM by enabling communication between child meshes for each of the four regions of the domain (i.e. interconnect, membrane assembly, air and fuel). In the computational code, the sub-meshes are included as 'fvMesh' (i.e. airMesh, fuelMesh, interconnectMesh, and electrolyteMesh), all of which consist of complete information of OpenFOAM computational meshes including, 'points', 'faces, 'owner', 'neighbour', 'boundary', and 'cellZones'. It does not mean a 'parent'/global mesh is not needed, since it represents an assembly of all sub-meshes, see Fig. 3. An OpenFOAM utility, "subsetMesh", is applied to define and extract each specific sub-mesh within the parent mesh. It should be noted that some faces which were previously internal to the parent mesh are now considered external faces of a child mesh and those faces are added to a patch named as "oldInternalFaces" by default, where the corresponding boundary conditions are set.

The sub-meshes in this model can be classified into three types, fluid, pure solid, and solid with electrochemical reactions. For instance, 'air' and 'fuel' belongs to the fluid type, 'interconnect' is solid, and the electrochemical reactions take place on the solid 'electrolyte'. In case oil cooling is necessary, as for the HT-PEFC stacks, an additional fluid region, 'oil', is implemented. The 'oil' region is simpler compared with the 'air' sub-mesh as although heat transfer occurs species transport is not needed.

Interphase heat and mass transfer rely on an important feature in OpenFOAM—'cellZones'. It represents subsets of mesh cells, which are lists of cell indexes for each zone. In the code, once the cellZones are defined using subsetMesh, communication can occur between each relevant pair of sub-meshes, e.g., between air and interconnect. For example, to apply the inter-phase transfer term for temperature, i.e. $\alpha_{\mathrm{j}}(T - T_{\mathrm{j}})$, between air ('airMesh') and interconnect ('interconnectMesh') on the zone named as 'active', the following lines of code are applied.

**Code example**

```
{
        volScalarField TairInterconnect(Tair * 0.0);    // Get the interconnect temperature

        //- Get the cell IDs for airMesh
        label znIdAir = airMesh.cellZones().findZoneID('active');
        labelList znAir = airMesh,cellZones()[znIdAir];

        //- Get the cell IDs for interconnectMesh
        label znIdInterconnect = interconnectMesh.cellZones().findZoneID('active');
        labelList znInterconnect = interconnectMesh.cellZones()[znIdInterconnect];

        forAll(znAir, cellI)
        {
                TairInterconnect[znAir[cellI]] = Tinterconnect[znInterconnect[cellI]];
        }
}
```

**Notes:**

- There exists an assumption that the orders and numbers of cells in 'active' zones are identical for 'airMesh' and 'interconnectMesh'. It is true in the case of extracting the sub-meshes from an identical global mesh, as in this work. If the sub-meshes are generated in a standalone manner, the code can predict unrealistic results or crash during run time.

  An alternative approach refers to the meshToMesh module in OpenFOAM. It offers an easier way to map fields between different sub-meshes.

- There are two widely used mesh manipulation tools in OpenFOAM, subsetMesh and splitMeshRegions. In the present model implementation, the former is applied rather than the latter. The former can extract any arbitrary zones from a 'polyMesh', without retaining the mapping information between the 'parent' and 'child' meshes or taking care of the interfacial boundaries between 'child' meshes. However, the latter can only split zones that are not overlapped with each other, which does not work in the present case.

Tip: In order to function better, 'topoSet' is usually needed before conducting subsetMesh.

The structure of model implementation can be found in Fig. 4 with filenames relevant to SOFCs, though it is similar for PEFCs. As in conventional OpenFOAM codes, the main file, sofcFoam.C, calls branches of header files (shown in in alphabetic order) making up the source code. The names of these files, e.g. 'createMesh', 'createFields', 'solveElectrochemistry', etc., indicate the solution process, which is similar to Beale et al. (2016), used to solve for the heat, species and momentum transport, variable gas and solid properties (e.g. density and thermal conductivity) and electrochemistry (e.g. current density distributions). It should be noted that the names of each sub-mesh are hardcoded, which means all the sub-meshes, 'air', 'fuel', 'electrolyte', 'interconnect', and/or 'oil', should be present when running the code.

**Fig. 4** The main structure of model implementation

**Code example (constant/air/porousZones)**

```
straight
{
    type            DarcyForchheimer; // porous Model
    active          yes;              // status
    cellZone        straight;         // name of porous zone
    porosity        0.07518797;       // porosity

    DarcyForchheimerCoeffs            // coefficents for DarcyForchheimer
    {
        d   (2.8e7 -1000 -1000);       // Darcy's parametr, d = 1/K
        f   (0 0 0);

        coordinateSystem
        {
            type    cartesian;
            origin  (0 0 0);
            coordinateRotation
            {
                type    axesRotation;
                e1      (1 0 0);    //(0.70710678 0.70710678 0);
                e2      (0 0 1);
            }
        }
    }

// ----------------------- diffusivity model ---------------------------//
        diffusivity
        {
            type        binaryFSG;
            Tname       T;
            pName       p;
            speciesA    O2;
            speciesB    N2;
        }
}
```

The solver enables the ‘runTimeSelectionTable’ option to be used for solving specific equations determined at the time of running the code. For example, two classes, ‘diffusivityModel’ and ‘porousZones’, are declared and instantiated in such a way. The desired species diffusivity model may be selected from four types of diffusion scenarios, including, binary diffusion (binaryFSG), diffusion with fixed coefficients (fixedDiffusivity), Knudsen diffusion (Knudsen), and diffusion in porous material (porousFSG). The porous model accounts for three types of porous descriptions, for instance, Darcy-Forchheimer, fixedCoeff, and powerLaw. These models are applied on the zones of each sub-mesh in the locations previously defined using ‘subsetMesh’. Therefore, the OpenFOAM platform enables the control required to selectively apply the desired equations in specific regions. It should be noted that the drag force/wall friction can be described by the Darcy-Forchheimer porous model. The parameters can be calculated accordingly. The use of the diffusivity and porous models is described as an example.

| Solution processes |
|---|
| 1. Initialize meshes, constants, and other parameters. Specify initial fields and boundary conditions, physical properties, cell voltage or prescribed average current density. |
| 2. Compute air and fuel density and viscosity and solve for pressure and momentum equations according to the SIMPLE algorithm. |
| 3. Calculate diffusion coefficients and solve for species transfer. |
| 4. Compute Nernst potential, polarization resistance, current density. |
| 5. Calculate the wall values of species mass fractions (on the ultra-thin electrolyte interfaces). |
| 6. Calculate the interphase heat transfer coefficients between adjacent sub-meshes. |
| 7. Obtain the heat sources and solve the energy equations of each sub-mesh sequentially. |
| 8. Repeating the steps 2 – 7 until convergence is finally researched. For the potentiostatic problem the voltage is fixed, whereas for the galvanostatic formulation, the cell voltage must be adjusted until the computed mean current density is identical to the set value. |
| Tip: the solution process can also be seen in the file sofcFoam.C. |

This chapter describes one of the first high temperature fuel cell stack models in OpenFOAM and its implementation represents a framework to consider multiple, co-located regions for coupling transport equations and electrochemical-reactions. Due to the complexity of the model, efforts have been taken to improve the stability and robustness of the solver to improve the range of operating conditions for which stable solutions are achievable. The improvement focuses on the consideration of (1) potential and current (Navasa et al. 2019), and (2) solver structure (Zhang et al. 2021) with marked success. The methods of the DRA stack model have been proven (Beale and Zhubrin 2005) to perform properly in another CFD package, PHOENICS. However, the implementation still needs to be carefully verified and validated to ensure the solver works as required. These comparisons can be seen in the following sections.

## 3 Verification of Model by Comparison with Direct Numerical Method

Several assumptions are made in the DRA to simulate stacks with reasonable computation times compared to the DNM model. Though well-established relations are used such as the Darcy friction factor, Nusselt number heat transfer resistance, and Sherwood number species transport resistance, the validity of their assumptions and their implementation in any model must be verified by comparison with an established reference. The following sections discusses verification of the DRA implementation in OpenFOAM by comparison with the DNM.

### 3.1 HT-PEFC Model Verification

The DRA simulation results were compared with DNM simulation results and experimental data for the temperature and pressure distributions.

The comparison between the numerical simulation (DRA) and experimental measurement for the temperature distribution is shown in Fig. 5. The temperature variations are in good agreement and slight deviations can be found in the hot zones, where the DRA predicts 2–3 K higher temperature than the experimental data. The differences may result from the thermal contact resistance between each component in the stack. The thermal contact resistance is not considered in the present model though it may/should be in future studies. However, the maximum temperature variation in the short stack is within 10 K, which exerts only slight effects on the performance of the stack.

The pressure distributions on the cathode side were also compared, as is shown in Fig. 6. The results from two different numerical methods, DRA and DNM, are also presented. The overall pressure drops from the inlet to the outlet are almost identical with only slight differences in the local pressure distributions. The DNM resolves the geometrical information of the fuel channel, which is especially important in the serpentine regions, the details can be found here (Zhang et al. 2019). In these, the drag forces due to the serpentine flow paths are larger than those in straight flow paths. A correction factor of $\xi = 2.5$ for tortuosity is therefore applied. However, a constant coefficient may not be able to represent the complex flow pattern and a more complex analytical formulation may improve agreement. Nevertheless, considering the relatively slight difference between the results of the DRA and DNM, the constant coefficient applied in the serpentine regions of the stack is considered a reasonable approximation for engineering purposes.

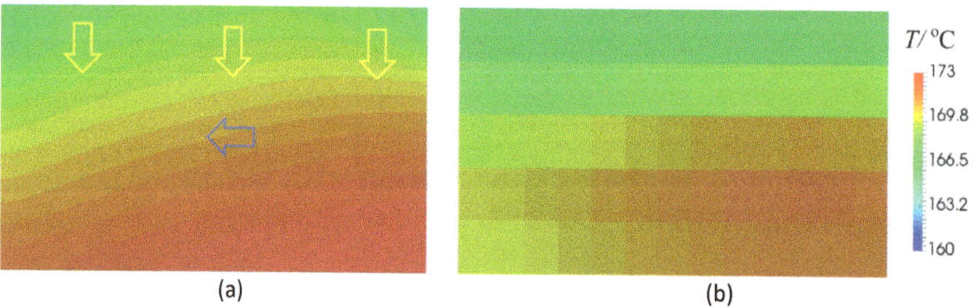


**Fig. 5** Temperature distribution on the interconnect for: **a** a model using DRA; and **b** experimental data (current density, $i = 0.6$ A cm$^{-2}$). Arrows, yellow: oil; blue: gases

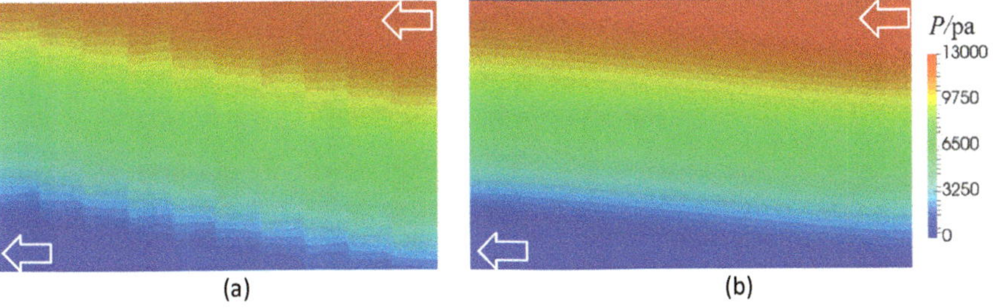


**Fig. 6** Modeled air side pressure distribution using: **a** a detailed numerical method (DNM); and **b** the DRA result (current density, $i = 0.6$ A cm$^{-2}$)

In PEFCs, the flow path designs vary greatly, in terms of spiral, pin, and inter-digitated, etc. In these applications, the averaging methods differ depending on the designs. Additional complexities include the thermal contact resistance, gas by-pass in the porous media, gas crossover through the electrolyte, and degradation, etc. These aspects are also important in future stack model developments.

### 3.1.1 SOFC Model Verification

For a single channel of an SOFC, Fig. 7a, b displays the current density distributions in the electrolyte for the DNM and the DRA model in counter-flow configurations, respectively. The air flows from the right to the left side, and fuel from the left to the right side. The maximum current densities can be observed at the fuel inlet, with the minima at the outlet. It can be seen that the local current density distributions are non-uniform normal to the main flow direction, with maximum values under the channel for the DNM, but uniform for the DRA stack model in the normal direction. The DNM and DRA presented good agreement for the current density distributions. Figure 7c, d shows the temperature distributions in the electrolytes in the counter-flow configurations for the DNM and the DRA model, respectively. The distributions were dominated by the air flows at the air inlets, where minimum values were detected. The temperatures also increased along the channels due to the exothermic electrochemical reactions. The present DRA model predicted slightly higher temperatures compared to the DNM. The good agreement between the DNM and DRA (as discussed further in Nishida et al. (2016)) demonstrate that the assumptions made in the DRA model to reduce computation time are valid at the appropriate length scales enabling extension to fuel cell stacks.

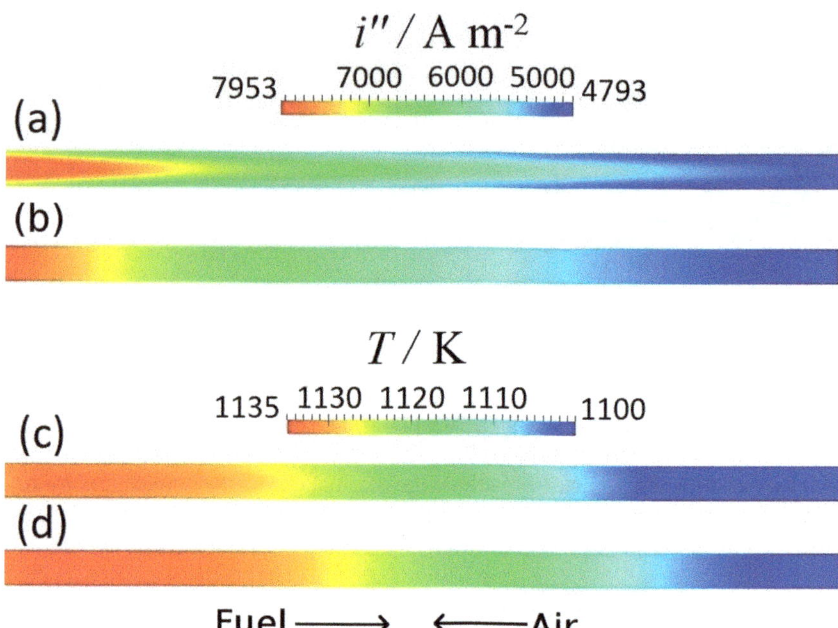


**Fig. 7** Current density, $i''$ (A m$^{-2}$) distribution in the PEN of SOFC for **a** detailed numerical model and **b** DRA model and temperature (K) in the PEN of SOFC for **c** detailed numerical model and **d** DRA model. SOFC simulated with mean current density of 6000 A m$^{-2}$ in counter-flow configuration

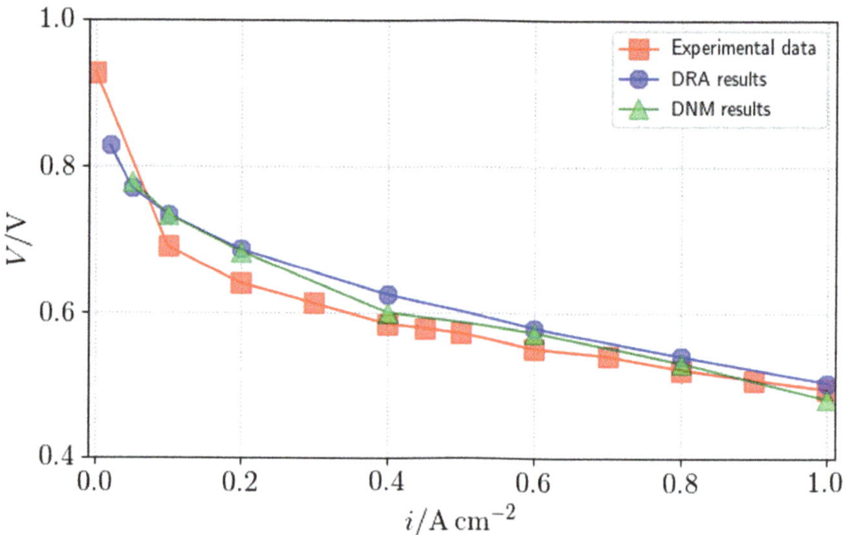


**Fig. 8** Comparison of polarization curves for numerical results and experimental data for a PEFC five-cell short stack

# 4 Validation of High Temperature Stack Models by Comparison with Experimental Results

In this section, the DRA model was applied to the in-house designed fuel cell stacks, as is shown in Fig. 1. The numerical predictions were compared with experimental measurements which were performed at Forschungszentrum Jülich, Germany. In addition, the simulation results offered the possibilities of obtaining the distributions of internal/local parameters that are usually unreachable using the current experimental apparatus, providing a useful tool for insights into fuel cell stack operation.

## *4.1 Experimental Validation of the HT-PEFC Short Stack Model*

The overall behavior of the short stack, represented by polarization curves, is numerically simulated and measured experimentally, as is depicted in Fig. 8. It can be seen that the differences between the predicted and measured voltages are slight, with a maximum deviation of 36 mV ($i = 0.2$ A cm$^{-2}$). The relative overall differences between the three different results are within 5%, which is minor considering the uncertainty that arises during experimental measurements. Therefore, the DRA is able to predict the overall performance of this stack. Additional comparisons are conducted on the local properties of the current density distributions.

Figure 9 displays the local current density distributions that were obtained numerically and experimentally. The local current density measurement was conducted by placing an S + + current shunt (Splusplus, www.splusplus.com") between the BPPs of the third and fourth cells (Kvesic, et al. 2012). The current density distributions decrease from the top-right to the bottom-left in all the cases. These results are in very good agreement while differences may be attributed to the resolutions of each method: the experimental data provide the lowest resolution, whereas the DNM presents the highest. The local current density distributions generally follow the oxygen molar fraction distribution.

By using an S + + current shunt device, the local temperature and current density distributions are measured, although with low resolution. However, the device is typically located at the outer surface of BBPs in order to avoid breaking the integrity of unit cells. The real values of temperature and current density inside the CCM/PEN are smeared through the BPPs, which means these values may not be readily captured. In addition, the gases' molar fraction/concentration cannot be directly measured locally. Therefore, the DRA or other mathematical models serve as important tools for enhancing the design and diagnosis of fuel cell stacks.

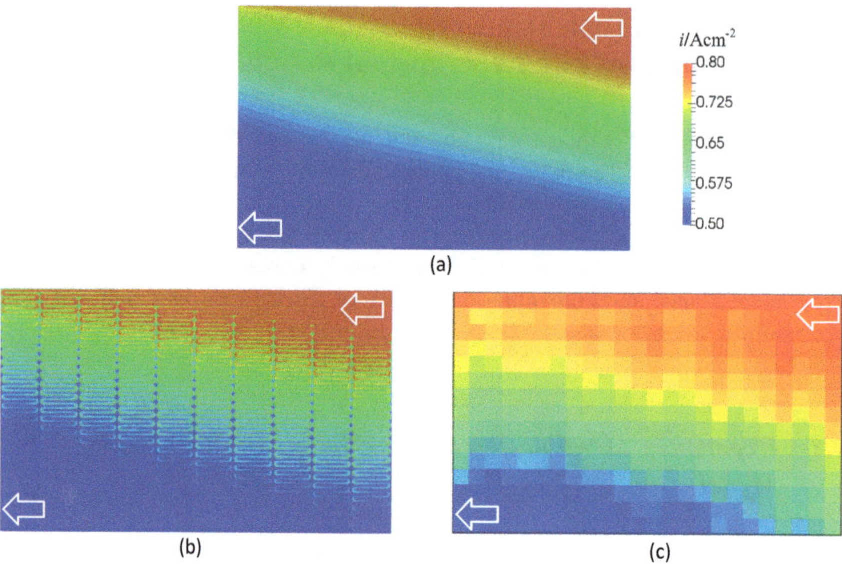


**Fig. 9** Current density distribution for: **a** DRA, **b** DNM, and (**c**). Experimental data, measured with S++, ,for mean current density, $i = 0.6$ A cm$^{-2}$, reproduced from Kvesic et al. (2012). (The arrows represent the flow directions)

## 4.2 *Experimental Validation of the Model of an 18-Cell Solid Oxide Fuel Cell Stack*

The performance of the 18-cell Jülich Mark-F SOFC stack was experimentally measured to obtain the polarization curve and power generation for given operating conditions. Under the same conditions, CFD simulations were conducted with the DRA stack model. The comparisons of various parameters are shown in Fig. 10. The results agree well at low current densities with some deviations at higher current densities. The numerical predictions also exhibited higher cell voltages and smaller deviations between each cell than the experimental measurements. The maximum differences of power output and local temperatures were within 0.25 kW and 50 °C, respectively.

It was indicated that the present stack model was able to qualitatively predict various parameters, with some quantifiable disagreement. On the one hand, the differences could result from the complex thermal–mechanical-electrical interactions and the potential of local changes in the physical properties due to stack degradation and fabrication errors, which are beyond the scope of the present model. On the other hand, a semi-empirical ASR function for electrochemical reactions was applied to describe the electrical resistance, which also introduced the opportunity for deviations as discussed in Nishida et al. (2018).

Figure 10b shows a comparison of local temperature variations between the numerical predictions and experimental data. Both results agree well at the fuel

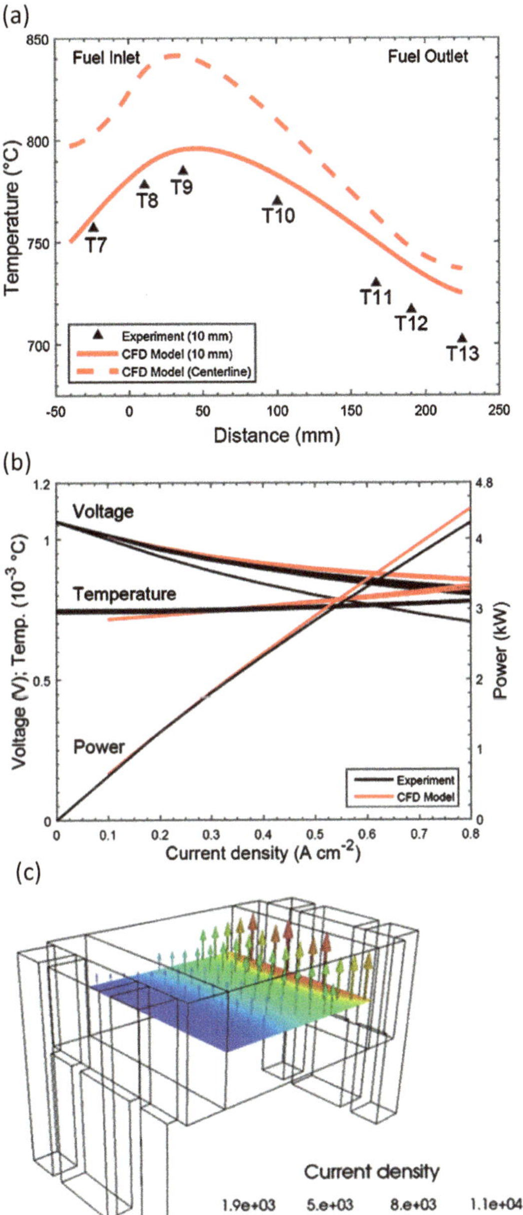


**Fig. 10** **a** Comparison of numerical predictions and experimental measurements for the power, temperature, and polarization curves of an 18-cell Jülich Mark-F SOFC stack, **b** comparisons of local temperature variations for the numerical results and experimental data, and **c** visualization of localized current density (A m$^{-2}$)

inlet, but present significant differences near the outlet. Under furnace conditions, the present model exhibits a higher overall temperature variation and a maximum difference of around 25 K, which is located near the fuel outlet. This may result from heat loss in the experimental set-up which was unaccounted for in the model. Nevertheless, the comparison provides confidence for the temperature at the center of the stack as predicted by the present model; locations which are typically inaccessible by experimental techniques. The maximum temperature along the central line was approximately 45 °C higher than the temperatures near the outer boundary. These temperature differences affect localized current density within each cell which in turn affects stack performance and so on, feedback which is captured by the model. For example, Fig. 15c shows the localized current density within the stack captured with the model which directly represents the Mark-F 18-cell solid oxide fuel cell stack of Fig. 1. The agreement between the experiment and model provides validation that the implementation in OpenFOAM can be used to provide insights into fuel cell operation and, considering the reasonable computation times of the DRA model, could be used in the design process.

## 5 Discussion and Summary

Modelling fuel cell stacks requires accurately capturing physical phenomena from the sub-micron to meter scales. However, due to computational limitations, it is necessary to make assumptions to reduce computational effort. This chapter describes the implementation of a Distributed Resistance Analogy (DRA) in OpenFOAM in which enables modeling of complex high temperature fuel cell stacks in reasonable computation times by making assumptions in the transport equations in select regions of the stack.

The DRA provides significant advantages for stack-level simulations, especially for relatively large and/or tall stacks. Though the local details are not resolved due to the assumptions of the method, those local details are better suited to a cell-scale model. Conversely, the DNM accounts for the full geometrical information and local momentum, heat and species diffusion; therefore, it can resolve local variations, however it consumes a large amount of computational time and hardware memory. For example, a five-cell (each cell: 200 $cm^2$) HT-PEFC short stack used in the DNM was meshed by tessellating the domain with approximately $172 \times 106$ of finite volume cells and required 36 h on a 900 core supercomputer demonstrating that the DNM has would have limited applicability in tall stacks of 100 s of cells. The advantages also apply to SOFC stacks, with a reduction in computation time of two orders of magnitude.

For fuel cells, the mass, momentum species and heat are transferred within and between different solid and fluid regions, and are in turn coupled with electrochemical reactions. The DRA employs a first order rate term (inter-phase transfer) which supplants the second order diffusion term in the transport equations in select regions to provide volume averaging where fine resolution is not necessary. The implementation of the DRA requires communication between cell center values between different solid and fluid regions of the stack. Since the implementation of the DRA method in OpenFOAM is unique to this work, it described in detail. Specifically, the transport equations are solved on distinct, but co-located meshes defined by the subsetMesh function from which cell center values are accessed. Within an overall solution procedure for coupled transport and electrochemistry, OpenFOAM provides the control required to DRA model to accurately represent real fuel cell stacks.

In this chapter, calculations for a short HT-PEFC stack were conducted with both a conventional DNM and a DRA model and both were validated with the detailed experimental results. Stack performance can be readily predicted using both numerical methods. The maximum deviations regarding the voltage-current relation were minor, and the local current density distributions were in good agreement between the numerical predictions and experimental measurements.

For the SOFC stack, voltage and temperature distributions predicted by the stack model (using the DRA) were compared with those from an 18-cell SOFC stack. The comparison indicated good overall agreement of the temperature distributions, with a maximum deviation of 3.3%. The stack model presented power curves close to the experimental data, but slightly over-predicted the local cell voltages and temperatures. The higher temperature and lower cell voltages may have resulted from the heat loss in the experimental set-up, and the transient effects during the ramp-up of the current in the experiment, respectively.

This work described implementation, verification and experimental validation of a comprehensive CFD model that coupled electrochemical reactions, fluid flow, and heat and mass transfer. Overall, the DRA model employed herein may be considered a good tradeoff between fidelity/granularity and computational effort. Its unique implementation in OpenFOAM enables accurate modelling of large-scale fuel cell stacks with the functionality, flexibility and control the OpenFOAM toolbox provides.

**Acknowledgements** The authors appreciate the help from many colleagues from IEK-14, Juelich Research Center and Queen's University. The support of Juelich Aachen Research Alliance (JARA) High Performance Computing (HPC) is important to conduct the numerical simulations. The language review and editing performed by Mr. Christopher Wood is also appreciated.

## References

Achenbach E (1994) Three-dimensional and time-dependent simulation of a planar solid oxide fuel cell stack. J Power Sources 49(1–3):333–348

Andersson M et al (2016) A review of cell-scale multiphase flow modeling, including water management, in polymer electrolyte fuel cells. Appl Energy 180:757–778

Arsalis A, Nielsen MP, Kær SK (2011a) Modeling and parametric study of a 1kWe HT-PEMFC-based residential micro-CHP system. Int J Hydrogen Energy 36:5010–5020

Arsalis A, Nielsen MP, Kær SK (2011b) Modeling and off-design performance of a 1kWe HT-PEMFC (high temperature-proton exchange membrane fuel cell)-based residential micro-CHP (combined-heat-and-power) system for Danish single-family households. Energy 36:993–1002
Beale SB (2015) Mass transfer formulation for polymer electrolyte membrane fuel cell cathode. Int J Hydrogen Energy 40:11641–11650
Beale SB et al (2016) Open-source computational model of a solid oxide fuel cell. Comput Phys Commun 200:15–26
Beale SB, Zhubrin SV (2005) A distributed resistance analogy for solid oxide fuel cells. Numer Heat Transf Part B: Fundam 47:573–591
Costamagna P et al (1994) Fluid dynamic study of fuel cell devices: simulation and experimental validation. J Power Sources 52(2):243–249
Hirschenhofer J, Stauffer D, Engleman R (1994) Fuel cell handbook. DIANE Publishing
Janßen H et al (2013) Development of HT-PEFC stacks in the kW range. Int J Hydrogen Energy 38:4705–4713
Janßen H et al (2017) Setup and experimental validation of a 5 kW HT-PEFC stack. Int J Hydrogen Energy 42(16):11596–11604
Janßen H et al (2018) Design and experimental validation of an HT-PEFC stack with metallic BPP. Int J Hydrogen Energy 43(39):18488–18497
Jeon DH et al (2010) Computational study of heat and mass transfer issues in solid oxide fuel cells
Kulikovsky AA (2019) Analytical modelling of fuel cells. Elsevier
Kvesic M et al (2012) 3D modeling of a 200 cm2 HT-PEFC short stack. Int J Hydrogen Energy 37:2430–2439
Kvesic M et al (2012) 3D modeling of a 200 cm2 HT-PEFC short stack. Int J Hydrogen Energy 37:2430--2439
Li Q et al (2009) High temperature proton exchange membranes based on polybenzimidazoles for fuel cells. Prog Polym Sci 34:449–477
Liu Z et al (2006) Numerical simulation of a mini PEMFC stack. J Power Sources 160(2):1111–1121
Liu Y et al (2016) A review of high-temperature polymer electrolyte membrane fuel-cell (HT-PEMFC)-based auxiliary power units for diesel-powered road vehicles. J Power Sources 311:91–102
Liu S et al (2019) Influence of operating conditions on the degradation mechanism in high-temperature polymer electrolyte fuel cells. J Power Sources 439: 227090
Lüke L et al (2012) Performance analysis of HT-PEFC stacks. Int J Hydrogen Energy 37:9171–9181
Nandjou F et al (2016a) A pseudo-3D model to investigate heat and water transport in large area PEM fuel cells – part 1: model development and validation. Int J Hydrogen Energy 41:15545–15561
Nandjou F et al (2016b) A pseudo-3D model to investigate heat and water transport in large area PEM fuel cells – Part 2: Application on an automotive driving cycle. Int J Hydrogen Energy 41:15545–15561
Navasa M, Miao X-Y, Frandsen HL (2019) A fully-homogenized multiphysics model for a reversible solid oxide cell stack. Int J Hydrogen Energy 44(41):23330–23347
Nishida R et al (2018) Three-dimensional computational fluid dynamics modelling and experimental validation of the Jülich Mark-F solid oxide fuel cell stack. J Power Sources 373:203–210
Nishida RT, Beale SB, Pharoah JG (2016) Comprehensive computational fluid dynamics model of solid oxide fuel cell stacks. Int J Hydrogen Energy 41:20592–20605
OpenFOAM, https://openfoam.org/.
Patankar SV, Spalding DB (1972) A calculation procedure for the transient and steady-state behaviour of shell-and-tube heat exchangers. Imperial College of Science and Technology, Department of Mechanical Engineering
Recknagle KP et al (2003) Three-dimensional thermo-fluid electrochemical modeling of planar SOFC stacks. J Power Sources 113(1):109–114
Roos M et al (2003) Efficient simulation of fuel cell stacks with the volume averaging method. J Power Sources 118(1):86–95

Shah RK, London AL (1978) Chapter VII - rectangular ducts. In: Shah RK, London AL (eds) Laminar flow forced convection in ducts. Academic Press, pp 196–222

Splusplus, *S++ scan shunt*. www.splusplus.com.

Steinberger-Wilckens R, Lehnert W (2010) Innovations in fuel cell technologies. Royal Society of Chemistry

Sun J et al (2018) A three-dimensional multi-phase numerical model of DMFC utilizing Eulerian-Eulerian model. Appl Therm Eng 132:140–153

Ubong EU, Shi Z, Wang X (2009) Three-dimensional modeling and experimental study of a high temperature PBI-based PEM fuel cell. J Electrochem Soc 156:B1276–B1282

Weber AZ et al (2014) A critical review of modeling transport phenomena in polymer-electrolyte fuel cells. J Electrochem Soc 161:F1254–F1299

Yuan P (2008) Effect of inlet flow maldistribution in the stacking direction on the performance of a solid oxide fuel cell stack. J Power Sources 185(1):381–391

Zhang S et al (2019) Modeling polymer electrolyte fuel cells: a high precision analysis. Appl Energy 233–234:1094–1103

Zhang S et al (2021) Modeling of reversible solid oxide cell stacks with an open-source library. ECS Trans 103(1):569–580

Zhang G, Jiao K (2018) Multi-phase models for water and thermal management of proton exchange membrane fuel cell: A review. J Power Sources 391:120–133

Zhang S, Beale SB, Reimer U, Andersson, M, & Lehnert W (2020) Polymer electrolyte fuel cell modeling - A comparison of two models with different levels of complexity. Int J Hydrog Energy 45(38): 19761–19777

# Effective Transport Properties

**Pablo A. García-Salaberri**

**Abstract** Porous media are an integral part of electrochemical energy conversion and storage devices, including fuel cells, electrolyzers, redox flow batteries and lithium-ion batteries, among others. The calculation of effective transport properties is required for designing more efficient components and for closing the formulation of macroscopic continuum models at the cell/stack level. In this chapter, OpenFOAM is used to determine the effective transport properties of virtually-generated fibrous gas diffusion layers. The analysis focuses on effective properties that rely on the fluid phase, diffusivity and permeability, which are determined by solving Laplace and Navier-Stokes equations at the pore scale, respectively. The model implementation (geometry generation, meshing, solver settings and postprocessing) is described, accompanied by a discussion of the main results. The dependence of orthotropic effective transport properties on porosity is examined and compared with traditional correlations.

## 1 Introduction

Macroscopic continuum models are based on a volume-averaged formulation of mass, momentum, species, charge and energy conservation equations (Weber et al. 2014; Wang 2004; Goshtasbi et al. 2019; García-Salaberri et al. 2017). The model is closed through appropriate constitutive relationships that define the various effective properties of the cell components (García-Salaberri et al. 2018). Effective transport properties include the absolute permeability used in Darcy's law, the tortuosity factor used to correct Fick's law of diffusion, or the effective electrical and ther-

**Supplementary Information** The online version contains supplementary material available at https://doi.org/10.1007/978-3-030-92178-1_3.

P. A. García-Salaberri (✉)
Departamento de Ingeniería Térmica y de Fluidos, Universidad Carlos III de Madrid, Leganés 28911, Spain
e-mail: pagsalab@ing.uc3m.es

© Springer Nature Switzerland AG 2022

S. Beale and W. Lehnert (eds.), *Electrochemical Cell Calculations with OpenFOAM*, Lecture Notes in Energy 42, https://doi.org/10.1007/978-3-030-92178-1_3

mal conductivities used in Ohm's and Fourier's laws. However, effective properties are challenging to determine in practice due to the thin nature of the porous components used in electrochemical devices (thickness $\sim 10 - 1000\ \mu$m), such as gas diffusion layers and catalyst layers in fuel cells and active electrodes in batteries (García-Salaberri et al. 2015a, b; Kashkooli et al. 2016; Liu et al. 2019). These porous media must fulfill several critical functions, such as providing a transport pathway for reactants/products through their pore volume and ensuring charge and heat conduction through their solid structure. Catalyst layers and active electrodes have the added functionality of providing a reactive surface area. Therefore, as a complement to experimentation, numerical simulation at the pore scale has become increasingly common. Pore-scale simulations in porous media provide direct insight into the impact of the microstructure on transport processes, allowing one to determine effective transport properties and explore specific transport phenomena (see, e.g., García-Salaberri et al. 2019; Hack et al. 2020; Sabharwal et al. 2016; Zhang et al. 2020; Gostick et al. 2007; Gostick 2013; Tranter et al. 2018; Belgacem et al. 2017; Aghighi and Gostick 2017). A thorough understanding of the mass, charge and heat transport properties of porous components is crucial for achieving improved performance and durability.

Two main pore-scale modeling approaches are widely used: pore-network modeling (PNM) and direct numerical simulation (DNS) (Arvay et al. 2012). PNMs idealize the pore space as a network of pore bodies interconnected by throats, whose size and connectivity are determined from the microstructure of the porous media (Gostick et al. 2007; García-Salaberri 2021). Some authors have also presented dual networks that include both the solid phase and the standard fluid phase (Aghighi and Gostick 2017). Different transport processes can be simulated on the network, including capillary transport, convection, diffusion and heat conduction. In contrast, DNS solves the transport equations (e.g., species conservation or Navier-Stokes equations) in computational meshes generated on tomography images or virtually-generated microstructures of porous media. Numerical methods used to solve conservation equations at the pore scale include the lattice Boltzmann method (LBM) or more conventional techniques such as the finite-element (FEM) or finite-volume (FVM) methods. Unlike the LBM, higher convergence rates are achieved with the FEM or FVM using steady-state solvers, although the time invested in mesh generation can represent a significant portion of the overall simulation (García-Salaberri et al. 2015a, b; Liu et al. 2019; Sabharwal et al. 2016). DNS only requires the input of the bulk properties of the constituents of the material (e.g., the bulk diffusion coefficient for effective diffusivity or the kinematic viscosity for absolute permeability), providing direct insight into the impact of microstructure on transport. Hence, the information that can be potentially extracted from DNS is higher, even though the computational cost is significantly higher than PNM.

Previous works that used OpenFOAM to simulate pore-scale transport phenomena in porous components of electrochemical energy conversion and storage devices are reviewed below. The literature survey includes works focused both on polymer electrolyte fuel cells (PEFCs) and solid oxide fuel cells (SOFCs).

In terms of PEFCs, James et al. (2012) examined the effect of inhomogeneous assembly compression on the effective electrical/thermal conductivity and diffusivity in a commercial GDL (SGL SIGRACET 30BA). The microstructure was extracted by means of X-ray computed tomography, then triangulated using the marching-cubes algorithm, and finally converted into a volumetric mesh for simulation. The numerical results showed that assembly compression significantly affects the effective transport properties between the under-the-land and under-the-channel regions. In addition, a notable decrease of the effective gas diffusivity was found compared to that predicted by widely used correlations, such as Bruggeman correlation (Bruggeman 1935) and the random fibre model of Tomadakis and Sotirchos (1993). Pharoah et al. (2011) analyzed the effective electrical/ionic conductivity and gas diffusivity of catalyst layers as a function of the volume fraction of carbon/Pt, ionomer and fluid phases. The microstructure was virtually-generated using spherical particles. It was found that Knudsen numbers in the pore space varied between the transition regime and Knudsen regime, with higher pore radius leading to lower Knudsen number. In a subsequent work, Khakaz-Babol et al. (2012) studied the coupling between transport and electrochemical kinetics on microstructural representations of catalyst layers that were generated using a similar algorithm to that of Pharoah et al. (2011). Different Pt loadings were created by randomly exchanging Pt particles with carbon particles, so that the base geometries were identical for each Pt loading. The results showed that both the transport of protons and oxygen significantly affect performance, with increased local losses in the ionomer at reduced Pt loading.

Regarding SOFCs, Choi et al. (2009) analyzed the effective electrical and ionic conductivity and gas diffusivity (including Knudsen diffusion) of the anode and cathode electrodes. The microstructures were made of randomly distributed and overlapping spheres with particle size distributions that matched those of ceramic powders. The numerical results were compared against experimental data and theoretical correlations. Gunda et al. (2011) examined the effective transport properties (electrical conductivity and gas diffusivity) of ceramic lanthanum strontium manganite (LSM) electrodes, whose microstructures were acquired using dual-beam focused ion beam-scanning electron microscopy (FIB-SEM). The sensitivity of different image processing steps (threshold value, median filter radius, morphological operators, surface triangulation, etc.) was examined. In addition, the work showed that the effective transport properties determined by FIB-SEM reconstruction were more anisotropic than those determined by numerical reconstruction. Next, Choi et al (2011) presented a numerical framework for the computation of effective transport properties of SOFC porous electrodes from 3D reconstructions of the microstructure based on measured parameters, such as porosity and particle size distribution. Three different types of grids were considered: cartesian, octree, and body-fitted/cut-cell with successive levels of surface refinement. OpenFOAM was used to compute the effective transport properties in the three phases of the electrode (pore, electron and ion). The model, validated with results from random walk simulations, was used to investigate microstructures with monosized particle distributions, as well as polydisperse particle size distributions similar to those found in SOFC electrodes. Bertei et al. (2014) presented a modeling framework, based on random sequential-addition packing algorithms, for the particle-based reconstruction of SOFC infiltrated electrodes. Key parameters, such as the connected triple-phase boundary length, effective electrical conductivity and effective diffusivity, were evaluated on the reconstructed electrodes by using geometric analysis, FVM and random-walk methods. A parametric study showed that the critical loading (i.e., the percolation threshold) increases as the backbone porosity decreases and the nanoparticle diameter increases. Large triple-phase boundary length, specific surface area and good effective conductivity can be reached by infiltration, without detrimental effect on effective transport properties of the fluid phase.

In this chapter, DNS in porous media using OpenFOAM is introduced. Diffusion and convection are simulated in virtually-generated GDLs formed by a 2D arrangement of randomly-oriented fibres. The organization of the chapter is as follows. In Sect. 2, the physical model is presented, including the governing equations, boundary conditions and calculation of effective transport properties, namely, effective diffusivity and permeability. In Sect. 3, the model implementation is described with a focus on the geometry generation, meshing, solution procedure and post-processing. In Sect. 4, the results are discussed, including the effect of porosity on the computed orthotropic transport properties and a comparison with traditional correlations.

## 2 Physical Model

Diffusion and convection in the fluid phase of fibrous GDLs are examined. The conservation equations and the expressions used to determine the corresponding effective transport properties, effective diffusivity and absolute permeability, are presented below.

**Effective Transport Properties of Solid Phase**

Effective properties that rely on the solid phase (e.g., electrical conductivity) can be determined using a similar procedure to that presented in this chapter, but changing the phase of interest.

### 2.1 *Diffusion: Effective Diffusivity*

Species molar concentration $C$ is determined from the steady-state species conservation equation (i.e., Laplace's equation)

$$\nabla \cdot (-D \nabla C) = 0 \tag{1}$$

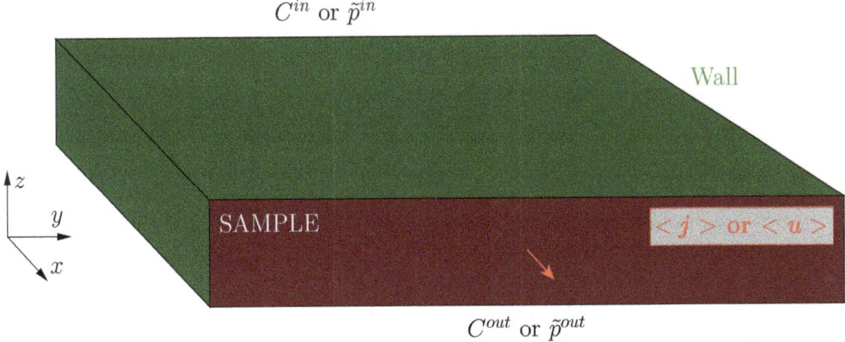


**Fig. 1** Schematic of the computational domain and boundary conditions used to determine effective transport properties in the in-plane direction ($x$-direction). Similar boundary conditions are applied to compute effective transport properties in other directions. Effective diffusivity and permeability are determined from the computed average diffusive flux and velocity in the direction of interest, respectively

where $D$ is the bulk diffusivity. This equation is subject to Dirichlet boundary conditions ($C = C^{\text{in}}$ and $C = C^{\text{out}}$) on the external faces of the domain perpendicular to the direction of interest $i$ ($i = x, y$ or $z$) to create a concentration gradient. $C^{\text{in}}$ and $C^{\text{out}}$ are the inlet and outlet concentrations, respectively. A no-flux boundary condition ($\partial C/\partial \boldsymbol{n} = 0$) is set on the remaining faces of the domain and internal fluid-solid interfaces. A schematic representation of the external boundary conditions is shown in Fig. 1.

According to Fick's first law, the effective diffusivity of an anisotropic porous medium is given by a second-order tensor, whose diagonal and non-diagonal components can be determined by changing the direction of interest in the calculations

$$\langle \boldsymbol{j} \rangle = -\bar{\bar{\boldsymbol{D}}}^{\text{eff}} \nabla C; \quad \bar{\bar{\boldsymbol{D}}}^{\text{eff}} = \begin{pmatrix} D^{\text{eff}}_{xx} & D^{\text{eff}}_{xy} & D^{\text{eff}}_{xz} \\ D^{\text{eff}}_{yx} & D^{\text{eff}}_{yy} & D^{\text{eff}}_{yz} \\ D^{\text{eff}}_{zx} & D^{\text{eff}}_{zy} & D^{\text{eff}}_{zz} \end{pmatrix} \tag{2}$$

where the symbol $\langle * \rangle$ denotes volume-average quantities. In this case, the volume-average diffusive flux.

Since the flux vanishes in the solid region ($s$) of the porous medium, i.e., $\boldsymbol{j}_s = 0$, the diagonal components of the normalized effective diffusivity tensor (e.g., the $zz$-component) are given by

$$\frac{D^{\text{eff}}_{ii}}{D} = \frac{\frac{1}{V_t}\int_{V_t} j_i dV}{\frac{D\Delta C}{L_i}} = \frac{\frac{V_f}{V_t}\left[\frac{1}{V_f}\int_{V_f} j_i dV\right]}{\frac{D\Delta C}{L_i}} = \varepsilon \frac{\langle j_{i,f} \rangle}{\frac{D\Delta C}{L_i}} \tag{3}$$

where $V_t$ and $V_f$ are the total and fluid ($f$) volume of the porous medium, respectively, $L_i$ is the length of the domain in the direction of interest $i$, $\varepsilon = V_f/V_t$ is the porosity, and $j_i = -D\partial_i C$ is the diffusive flux in $i$-direction. Non-diagonal components were not determined since they are typically small in the fibrous materials examined (see the note below).

**Non-Diagonal Components**

The full effective diffusivity tensor $\bar{\bar{D}}^{\text{eff}}$ can be determined from three simulations (1,2,3) of diffusive flux fields $(j_x^1, j_y^1, j_z^1)$, $(j_x^2, j_y^2, j_z^2)$ and $(j_x^3, j_y^3, j_z^3)$, corresponding to imposed concentration gradients in the $x$-, $y$- and $z$-direction, $\nabla C_x^1$, $\nabla C_y^2$ and $\nabla C_z^3$, respectively.
Using Fick's first law, the components of the tensor are obtained by solving the following system of equations

$$-\begin{pmatrix} D_{xx}^{\text{eff}} & D_{xy}^{\text{eff}} & D_{xz}^{\text{eff}} \\ D_{yx}^{\text{eff}} & D_{yy}^{\text{eff}} & D_{yz}^{\text{eff}} \\ D_{zx}^{\text{eff}} & D_{zy}^{\text{eff}} & D_{zz}^{\text{eff}} \end{pmatrix} \begin{pmatrix} \nabla C_x^1 & \nabla C_x^2 & \nabla C_x^3 \\ \nabla C_y^1 & \nabla C_y^2 & \nabla C_y^3 \\ \nabla C_z^1 & \nabla C_z^2 & \nabla C_z^3 \end{pmatrix} = \begin{pmatrix} \langle j_x^1 \rangle & \langle j_x^2 \rangle & \langle j_x^3 \rangle \\ \langle j_y^1 \rangle & \langle j_y^2 \rangle & \langle j_y^3 \rangle \\ \langle j_z^1 \rangle & \langle j_z^2 \rangle & \langle j_z^3 \rangle \end{pmatrix} \tag{4}$$

where $\nabla C_x^2$, $\nabla C_x^3$, $\nabla C_y^1$, $\nabla C_y^3$, $\nabla C_z^1$ and $\nabla C_z^2$ are the average concentration gradients computed in the transverse directions.
Note that although a local no-flux boundary condition is prescribed at the sidewalls of the domain, the average concentration gradients and volume-average diffusive fluxes in the transverse directions are in general different from zero. Similar considerations apply for the permeability tensor (see Guibert et al. (2016) for further details).

## 2.2 *Convection: Permeability*

Convection is modeled through the steady-state mass conservation and Navier-Stokes equations for an incompressible Newtonian fluid

$$\begin{aligned} &\nabla \cdot \boldsymbol{u} = 0 \\ &\nabla \cdot (\boldsymbol{u}\boldsymbol{u}) = -\nabla \tilde{p} + \nabla \cdot (\nu\,(\nabla \boldsymbol{u} + \nabla \boldsymbol{u}^{\intercal})) \end{aligned} \tag{5}$$

where $\boldsymbol{u}$ is the velocity vector, $\nu$ is the kinematic viscosity, $p$ is the static pressure and $\tilde{p} = p/\rho$ is the kinematic pressure, with $\rho$ the density. Similar to diffusion, Dirichlet boundary conditions are prescribed for pressure on the external faces of the domain in $i$-direction, $\tilde{p} = \tilde{p}^{\text{in}}$ and $\tilde{p} = \tilde{p}^{\text{out}}$. An impermeable no-slip boundary condition ($\boldsymbol{u} = 0$) is set on the remaining faces of the domain and interior fluid-solid interfaces of the porous medium.

According to Darcy's law,

$$\langle \boldsymbol{u} \rangle = -\frac{\overline{\overline{\boldsymbol{K}}}}{\nu} \nabla \tilde{p}; \quad \overline{\overline{\boldsymbol{K}}} = \begin{pmatrix} K_{xx} & K_{xy} & K_{xz} \\ K_{yx} & K_{yy} & K_{yz} \\ K_{zx} & K_{zy} & K_{zz} \end{pmatrix}, \tag{6}$$

the diagonal components of the permeability tensor, $K_{ii}$, are determined as

$$K_{ii} = \frac{\nu \left[ \frac{1}{V_t} \int_{V_t} u_i dV \right]}{\frac{\Delta \tilde{p}}{L_i}} = \frac{\nu \frac{V_f}{V_t} \left[ \frac{1}{V_f} \int_{V_f} u_i dV \right]}{\frac{\Delta \tilde{p}}{L_i}} = \varepsilon \nu L_i \frac{\langle u_{if} \rangle}{\Delta \tilde{p}} \tag{7}$$

where $\Delta \tilde{p} = \tilde{p}^{\text{in}} - \tilde{p}^{\text{out}}$ is the prescribed kinematic pressure difference and $u_i$ is the $i$-component of the velocity vector.

Calculations of absolute permeability are usually performed in physical units, considering $\Delta \tilde{p} = 1\ \text{m}^2/\text{s}^2$ ($\tilde{p}^{\text{in}} = 1\ \text{m}^2/\text{s}^2$, $\tilde{p}^{\text{out}} = 0$), while $\nu$ is adjusted to ensure that the flow is in the creeping regime, i.e.,

$$Re = \frac{\langle \|\boldsymbol{u}\| \rangle d_f}{\nu} \approx \frac{\langle u_i \rangle d_f}{\nu} \ll 1 \tag{8}$$

where $\langle \|\boldsymbol{u}\| \rangle$ is the average modulus of the velocity and $d_f$ is the diameter of the mono-sized fibres. Using these values, the expression of permeability is reduced to

$$K_{i,i} = \varepsilon \nu L_i \langle u_{i,f} \rangle \tag{9}$$

where $\langle u_{i,f} \rangle$ is the average velocity in $i$-direction in the fluid phase.

**Darcy's Law**

Darcy's law can be verified by varying $\Delta \tilde{p}$, while keeping constant the value of $\nu$. The linear relationship between $\Delta \tilde{p}/L_i$ and $Re$ shows that inertia is not important and the flow is indeed in the creeping regime, $Re \ll 1$.

## 3 Model Implementation

The main steps followed for the calculation of the effective transport properties are presented in this section, including the geometry generation, meshing routine, solver selection and post-processing. An example of the bash script used to run the simulations of effective diffusivity (and permeability) is presented below. The number of processors in the parallel execution is given as an argument to the script. Before running the script, the user must generate the triangulated geometry of the porous medium (*facets.stl*) using the *gdl.cpp* code. In addition to the solution fields, the output results include the volume-average diffusive fluxes (or velocity components) in each direction, which are saved periodically (as indicated in *controlDict*) and in the last iteration. The computed values are written into the *postProcessing* folder. The average quantities corresponding to the last iteration are used to determine the effective transport properties through Eqs. (3) and (9). The dimensions of the domain, porosity and bulk properties are known in advance.

**Listing 3.1** Main script used to run simulations starting from the geometry file *facets.stl*.

```
#Argument: number of processors (equal to the number of
    subdomains in decomposeParDict)

#Clean folders
foamListTimes -rm
rm -r ./postProcessing
rm -r ./processor*

#########
#MESHING
#########

#Run blockMesh
blockMesh > log

#Copy files
cp facets.stl ./constant/triSurface

#Run snappyHexMesh
decomposePar >> log
mpirun -np $1 snappyHexMesh -overwrite -parallel >> log

#Set initial fields
ls -d processor* | xargs -I {} rm -rf ./{}/0
ls -d processor* | xargs -I {} cp -r 0.orig ./{}/0

#Scale mesh to meters [m]
mpirun -np $1 transformPoints -scale "(0.001 0.001 0.001)" -
    parallel >> log

#Renumber mesh
mpirun -np $1 renumberMesh -overwrite -parallel >> log

#Check final mesh
mpirun -np $1 checkMesh -parallel >> log

############
#RUN SOLVER
############
```

```
(diffusion) mpirun -np $1 laplacianFoam -parallel -
    noFunctionObjects >> log
(convection) mpirun -np $1 simpleFoam -parallel -
    noFunctionObjects >> log

################
#POSTPROCESSING
################

#Create cell region
mpirun -np $1 topoSet -parallel >> log

#Calculate average fluxes/velocities
(diffusion) mpirun -np $1 postProcess -fields "(T gradTx gradTy
    gradTz)" -parallel >> log
(convection) mpirun -np $1 postProcess -fields "(U p)" -
parallel >> log
```

**Managing Computational Simulations**

Since the meshing step is time consuming, it is recommended to use the decomposed mesh for calculations in other directions. The boundary conditions (i.e., the direction of the imposed gradient) must be changed as desired in the *0.orig* folder.

## 3.1 Geometry

The microstructure of porous media can be obtained from (1) tomography images (García-Salaberri et al. 2015a, b, 2018, 2019) or can be (2) generated virtually using numerical algorithms (Choi et al. 2009; Choi et al 2011; Bertei et al. 2014). Volume-averaged quantities (e.g., composition fraction) and statistical descriptors (e.g., pore size distribution, n-point correlation functions, lineal path function, and chord length function) can be used as objective variables (Pant et al. 2014). Usually, the first method provides a higher degree of fidelity to reality, even though sometimes it is difficult to determine the constituents present in the images. An example is the differentiation of carbon fibres/binder and polytetrafluoroethylene (PTFE) in GDLs due to their similar X-ray absorption properties García-Salaberri et al. (2018). The virtual generation of materials allows one to overcome this issue, although the creation of realistic microstructures can be challenging in some circumstances. An example is the complex multi-component, multi-scale geometry of catalyst layers. Here, fibrous porous materials with a 2D arrangement of randomly-oriented fibres similar to carbon-paper GDLs were used as an illustrative example.

For a specified number of cylindrical fibres, $N_f$, of diameter $d_f = 10\ \mu$m, the steps followed for the generation of the material microstructure in a box of size $[0 - S_x, 0 - S_y, 0 - S_z]$ ($S_x = S_y = 1.5$ mm, $S_z = 0.25$ mm), are as follows:

1. The 3D coordinates, $\boldsymbol{P_1}$ and $\boldsymbol{P_2}$, of the axial endpoints of the cylindrical fibres are generated randomly inside the box of material. The same $z$-coordinate is prescribed for the endpoints of each fibre ($P_{1,z} = P_{2,z}$) to achieve a 2D arrangement.
2. The $x$ and $y$ coordinates of the axial endpoints are translated $-0.25$ mm, so there is some extra material around the cropped domain of size $[0 - L_x, 0 - L_y, 0 - L_z]$ that is used in the simulations ($L_x = L_y = 1$ mm, $L_z = 0.25$ mm). This step removes edge effects.
3. Once the position of all the axial endpoints is fixed, the lateral surface of the cylindrical fibres is triangulated using a spacing in the axial and azimuthal directions, $\boldsymbol{\Delta x} = (\boldsymbol{P_2} - \boldsymbol{P_1})/10$ and $\Delta\phi = 2\pi/20$, respectively. The unit normal vectors perpendicular to each triangle are also determined.
4. The vertices and normals of the triangles that define the lateral surface of the cylindrical fibres are written into an STL file (*facets.stl*).
5. The endcaps of the cylindrical fibres are triangulated using a spacing in the azimuthal direction equal to that used for the lateral surface ($\Delta\phi = 2\pi/20$).
6. The vertices and normals of the triangles that define the endcaps are added to the *facets.stl* file (an example of a triangulated geometry is shown in Fig. 2).

**Fibre Intersections**

Fibre intersection is not explicitly taken into account in the geometry generation. However, when the fluid region is selected in the meshing step with *snappyHexMesh*, the mesh is adapted to the external fibres surface and the intersections among them. The solid regions inside the fibres and their intersections are removed, since they are unreachable from the fluid region.

## 3.2 Meshing

The meshing utility *snappyHexMesh* is used to mesh the pore space enclosed within the external surfaces of the domain and the geometry of the cylindrical fibres. A background mesh composed of cubes is first created with *blockMesh*. Then, the resulting mesh is refined with *snappyHexMesh* using the *facets.stl* file as an input. Castellated meshes were considered here, suppressing the surface snapping and layer addition steps. The resulting meshes had around 5.5 millions of cells depending on the number of fibres $N_f$ in the sample (i.e., the porosity). The maximum number of cells was achieved for intermediate porosities around $\varepsilon \approx 0.6$. An example of the generated meshes is shown in Fig. 3, including a close-up view of the refined mesh close to the fibres surface. The level of refinement used in *snappyHexMesh* was set equal to (2 4).

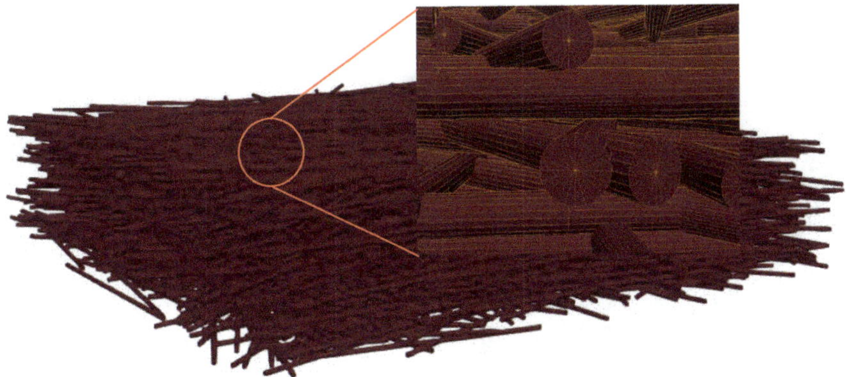

**Fig. 2** Geometry composed of cylindrical fibres of 10 $\mu$m in diameter with random orientations in the material plane ($x$-$y$ plane) used to mimic the microstructure of binder-free carbon-paper GDLs. The close-up view shows the triangulated geometry

Practical Advice

The following guidelines should be taken into account during the meshing step:

1. Introduce a small gap slightly higher than the fibre radius near the boundary faces in the $z$-direction, so that the axial endpoints of the fibres do not touch the boundary faces. This practice facilitates the selection of a point inside the fluid region during the meshing step.
2. Set the size of the cubic cells in the background mesh similar to the fibre diameter ($d_f = 10\ \mu$m). This size ensures that the geometry of the fibres is properly captured during the refinement of the mesh with *snappyHexMesh*. No significant differences were found in the results using background cells of 5 $\mu$m.

## 3.3 Solver and Post-processing

Laplace and Navier-Stokes equations used to simulate diffusion and convection are solved in OpenFOAM with the steady-state solvers *laplacianFoam* and *simpleFoam*, respectively. A laminar *simulationType* with a Newtonian *transportModel* is used in *simpleFoam*. The remaining solver settings can be kept similar to those commonly used in OpenFOAM.

For post-processing, the *topoSet* utility is used to create a cell zone that includes the entire computational domain (i.e., the fluid region of the porous medium). Then,

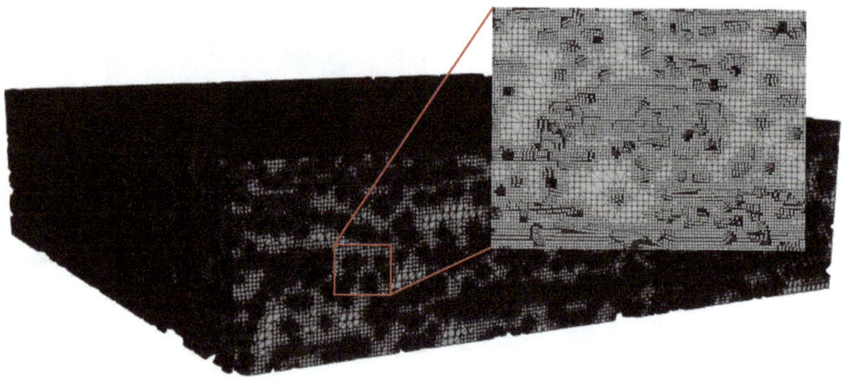

**Fig. 3** Castellated mesh generated with *snappyHexMesh* using a background mesh with cubic cells of 10 $\mu$m in size (equal to the fibre diameter, $d_f$). The close-up view shows the refined mesh around the fibres surface

the volume-average diffusive fluxes (or velocity components) in each direction are calculated using the *postProcess* utility, defining function objects in *controlDict*. The *volAverage* operation is used for volume averaging.

**Calculation of Effective Transport Properties**

The pre-factors multiplying the average diffusive flux (or velocity component) in Eqs. (3) and (9) (such as the length $L_i$ or the porosity $\varepsilon$) can be included as a scaling factor in the function objects implemented in *controlDict*. The porosity can be easily determined by dividing the volume of the fluid region (provided by the *postProcess* utility) by the total volume of the domain (which is known in advance). Alternatively, the pre-factors can be introduced when the results are plotted, for example, using Python.

## 4 Results

In this section, the computed results for the effective diffusivity and permeability are discussed. The results in both the through- and in-plane directions are presented given the anisotropy of the generated carbon-paper GDLs. One representative direction ($x$-direction) is analyzed in the material plane ($x$-$y$ plane). Figures 4 and 5 show some illustrative examples of the species concentration distributions corresponding to effective diffusivity calculations for two different porosities (i.e., number of fibres). The pressure distributions and streamlines corresponding to permeability calculations are shown in Figs. 6 and 7. The porosities are in the range typically

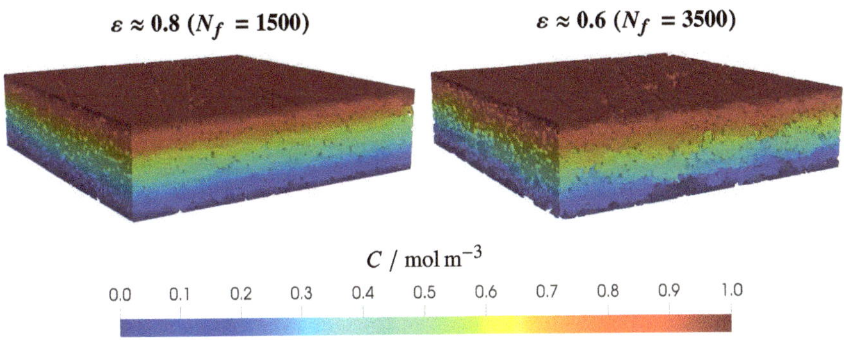


**Fig. 4** Concentration fields, $C(x, y, z)$, corresponding to calculations of the through-plane effective diffusivity for two different porosities, $\varepsilon$ (number of fibres, $N_f$): (left) $\varepsilon = 0.8$ ($N_f = 1500$), (right) $\varepsilon = 0.6$ ($N_f = 3500$)

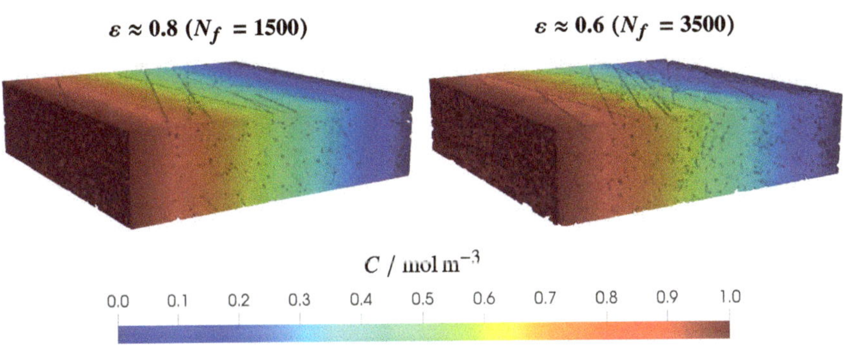


**Fig. 5** Concentration fields, $C(x, y, z)$, corresponding to calculations of the in-plane effective diffusivity. See caption to Fig. 4 for further details

observed for uncompressed ($\varepsilon \approx 0.8$) and mid-compressed ($\varepsilon \approx 0.6$) GDLs. Higher transport properties are found in the in-plane direction due to the 2D arrangement of carbon fibres (and pores), which facilitates transport in this direction.

The overall diffusive flux across the material (at the same concentration gradient) decreases with decreasing porosity and increasing tortuosity of transport pathways. The increment of tortuosity with porosity leads to non-linearities in the dependence between the normalized effective diffusivity, $D^{\text{eff}}/D$, and the porosity, $\varepsilon$, as given by the relationship García-Salaberri et al. (2015b):

$$\frac{D^{\text{eff}}}{D} = \frac{\varepsilon}{\tau} \tag{10}$$

where $\tau$ is the diffusion tortuosity factor. For example, $\tau = \varepsilon^{-1/2}$ ($D^{\text{eff}}/D = \varepsilon^{1.5}$) in the traditional Bruggeman correction for porous media consisting of small, spherical solid inclusions (Bruggeman 1935; Tjaden et al. 2016).

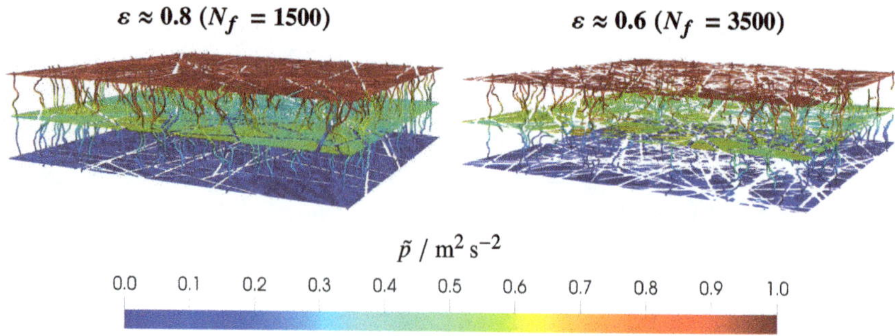


**Fig. 6** Kinematic pressure fields, $\tilde{p}(x, y, z)$, and streamlines (colored by pressure level) corresponding to calculations of the through-plane absolute permeability for two different porosities, $\varepsilon$ (number of fibres, $N_f$): (left) $\varepsilon = 0.8$ ($N_f = 1500$), (right) $\varepsilon = 0.6$ ($N_f = 3500$)

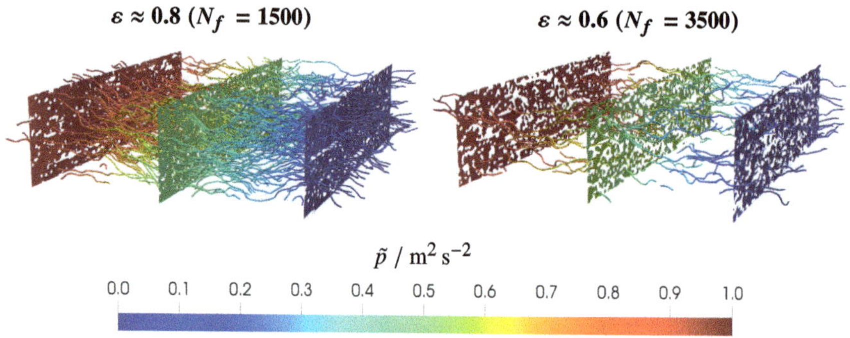


**Fig. 7** Kinematic pressure, $\tilde{p}(x, y, z)$, and streamlines (colored by pressure level) corresponding to calculations of the in-plane absolute permeability. See caption to Fig. 6 for further details

Permeability is influenced by hydraulic radius, porosity and tortuosity, resulting in a non-linear variation as a function of porosity Holzer et al. (2017). The Carman-Kozeny equation has been successfully applied to describe the variation of GDL permeability with porosity Gostick et al. (2006).

$$K = \frac{d_f^2 \varepsilon^3}{16 k_{ck} (1 - \varepsilon)^2} \tag{11}$$

where $k_{ck}$ is the Carman-Kozeny constant, which is used as a fitting parameter depending on the microstructure of the porous medium.

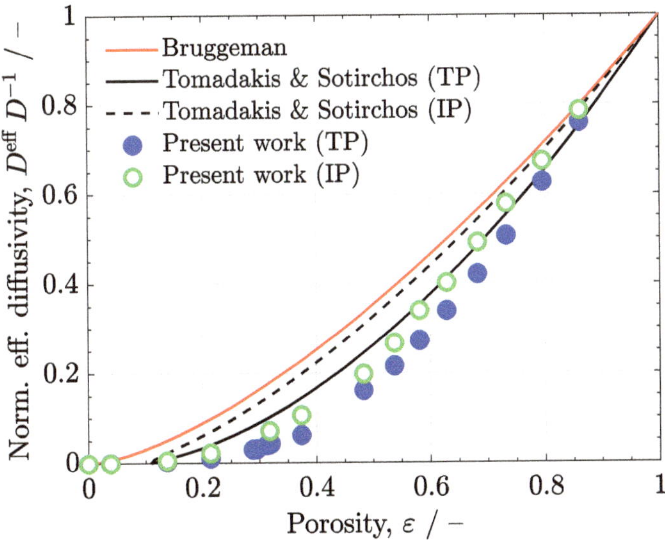


**Fig. 8** Computed through- (TP) and in-plane (IP) normalized effective diffusivities, $D^{\text{eff}}/D$, as a function of porosity, $\varepsilon$. The Bruggeman correlation Bruggeman (1935), $D^{\text{eff}}/D = \varepsilon^{1.5}$, and the anisotropic random fibre model of Tomadakis and Sotirchos (1993) (see Eq. (12)) are also shown

The variations of the normalized effective diffusivity and permeability as a function of porosity are shown in Figs. 8 and 9, respectively. The normalized effective diffusivity is lower than that predicted by the Bruggeman correlation, $D^{\text{eff}}/D = \varepsilon^{1.5}$, due to the more complex geometry of fibrous GDLs. Moreover, it is somewhat lower than the correlation proposed by Tomadakis and Sotirchos (1993) for random fibre structures

$$\frac{D^{\text{eff}}}{D} = \varepsilon \left( \frac{\varepsilon - 0.11}{1 - 0.11} \right)^{n} \tag{12}$$

where $n = 0.785$ and $n = 0.521$ for the through- and in-plane directions, respectively.

The differences between both models are ascribed to the different methodology used for the generation of the fibrous geometry. For instance, the values computed here approach those reported for binder-free GDLs, such as Freudenberg carbon paper, being $D^{\text{eff}}/D \approx 0.36$ for $\varepsilon \approx 0.65$ (Hack et al. 2020; Hwang and Weber 2012).

The permeability is well correlated as a function of porosity using Eq. (11) with $k_{ck} = 2 - 4$. A steeper decrease of the permeability is found for porosities below $\varepsilon \lesssim 0.5$, which drops around two orders of magnitude in the range $\varepsilon = 0.1 - 0.5$ compared to the ten-fold descent in the range $\varepsilon = 0.5 - 0.85$. The non-linear behaviour arises from the sensitivity of permeability to small microstructural differences near the percolation threshold. Similar results were reported in previous works for fibrous porous media (Tomadakis and Robertson 2005; Nabovati et al. 2009).

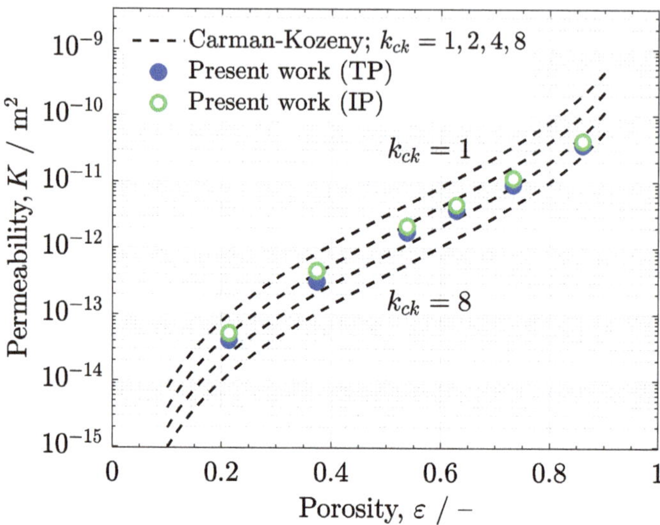


**Fig. 9** Computed through- (TP) and in-plane (IP) absolute permeabilities, $K$, as a function of porosity, $\varepsilon$. The curves corresponding to various Carman-Kozeny constants, $k_{ck}$, are also shown (see Eq. (11))

### Effective Diffusivity and Permeability of Commercial GDLs

For a given porosity, the effective diffusivity and permeability of commercial GDLs (Toray, SGL Carbon Group and Freudenberg carbon papers) highly depends on their microstructure. In fact, the volume fraction and porosity of binder have been identified as key parameters that influence the effective transport properties of GDLs (García-Salaberri et al. 2018; Mathias et al. 2003; Zenyuk et al. 2016). For example, Toray TGP-H series carbon papers show lower effective diffusivities in the through-plane direction than those computed here due to the more complex pore structure that arises from the addition of an almost non-porous binder (García-Salaberri et al. 2015a, b); $D^{\text{eff}}/D \sim 0.22 - 0.28$ (Toray TGP-H) vs. $D^{\text{eff}}/D \sim 0.45$ (present work) at $\varepsilon \approx 0.72$.

**Acknowledgements** This work was supported by the projects PID2019-106740RB-I00 and EIN2020-112247 (Spanish Agencia Estatal de Investigación) and the project PEM4ENERGY-CM-UC3M funded by the call "Programa de apoyo a la realización de proyectos interdisciplinares de I+D para jóvenes investigadores de la Universidad Carlos III de Madrid 2019-2020" under the frame of the "Convenio Plurianual Comunidad de Madrid-Universidad Carlos III de Madrid".

## References

Aghighi M, Gostick J (2017) Pore network modeling of phase change in PEM fuel cell fibrous cathode. J Appl Electrochem 47:1323–1338

Arvay A, Yli-Rantala E, Liu CH, Peng XH, Koski P, Cindrella L, Kauranen P, Wilde PM, Kannan AM (2012) Characterization techniques for gas diffusion layers for proton exchange membrane fuel cells: a review. J Power Sources 213:317–337

Belgacem N, Prat M, Pauchet J (2017) Coupled continuum and condensation-evaporation pore network model of the cathode in polymer-electrolyte fuel cell. Int J Hydrogen Energy 42:8150–8165

Bertei A, Pharoah JG, Gawel DAW, Nicolella C (2014) A particle-based model for effective properties in infiltrated solid oxide fuel cell electrodes. J Electrochem Soc 161:F1243–F1253

Bruggeman DAG (1935) Berechnung verschiedener physikalischer Konstanten von heterogenen Substanzen. I. Dielektrizitätskonstanten und Leitfähigkeiten der Mischkórper aus isotropen Substanzen. Ann Phys 5:636–664

Choi HW, Berson A, Pharoah JG, Beale S (2011) Effective transport properties of the porous electrodes in solid oxide fuel cells. Proc IMechE Vol Part A: J Power Energy 225:183—197

Choi HW, Berson A, Kenney B, Pharoah JG, Beale S, Karan K (2009) Effective transport coefficients for porous microstructures in solid oxide fuel cells. ECS Trans 25:1341–1350

García-Salaberri PA (2021) Modeling diffusion and convection in thin porous transport layers using a composite continuum-network model: Application to gas diffusion layers in polymer electrolyte fuel cells. Int J Heat Mass Transf 167:120824

García-Salaberri PA, Gostick JT, Hwang G, Weber AZ, Vera M (2015a) Effective diffusivity in partially-saturated carbon-fiber gas diffusion layers: effect of local saturation and application to macroscopic continuum models. J Power Sources 296:440–453

García Salaberri PA, Hwang G, Vera M, Weber AZ, Gostick JT (2015b) Effective diffusivity in partially-saturated carbon-fiber gas diffusion layers: effect of through-plane saturation distribution. Int J Heat Mass Transf 86:319–333

García-Salaberri PA, García-Sánchez D, Boillat P, Vera M, Friedrich KA (2017) Hydration and dehydration cycles in polymer electrolyte fuel cells operated with wet anode and dry cathode feed: a neutron imaging and modeling study. J Power Sources 359:634–655

García-Salaberri PA, Zenyuk IV, Shum AD, Hwang G, Vera M, Weber AZ, Gostick JT (2018) Analysis of representative elementary volume and through-plane regional characteristics of carbon-fiber papers: diffusivity, permeability and electrical/thermal conductivity. Int J Heat Mass Transf 127:687–703

García-Salaberri PA, Hwang G, Vera M, Weber AZ, Gostick JT (2019) Implications of inherent inhomogeneities in thin carbon fiber-based gas diffusion layers: a comparative modeling study. Electrochim Acta 295:861–874

Goshtasbi A, García-Salaberri PA, Chen J, Talukdar K, García-Sánchez D, Ersal T (2019) Through-the-membrane transient phenomena in PEM fuel cells: a modeling study. J Electrochem Soc 166:F3154–F3179

Gostick JT (2013) Random pore network modeling of fibrous PEMFC gas diffusion media using Voronoi and Delaunay Tessellations. J Electrochem Soc 160:F731–F743

Gostick JT, Fowler MW, Pritzker MD, Ioannidis MA, Behra LM (2006) In-plane and through-plane gas permeability of carbon fiber electrode backing layers. J Power Sources 162:228–238

Gostick JT, Ioannidis MA, Fowler MW, Pritzker MD (2007) Pore network modeling of fibrous gas diffusion layers for polymer electrolyte membrane fuel cells. J Power Sources 173:277–290

Guibert R, Horgue P, Debenest G, Quintard M (2016) A comparison of various methods for the numerical evaluation of porous media permeability tensors from pore-scale geometry. Math Geosci 48:329–347

Gunda NSK, Choi HW, Berson A, Kenney B, Karan K, Pharoah JG, Mitra SK (2011) Focused ion beam-scanning electron microscopy on solid-oxide fuel-cell electrode: image analysis and computing effective transport properties. J Power Sources 196:3592–3603

Hack J, García-Salaberri PA, Kok MDR, Jervis R, Shearing PR, Brandon N, Brett DJL (2020) X-ray micro-computed tomography of polymer electrolyte fuel cells: what is the representative elementary area? J Electrochem Soc 167:013545
Holzer L, Pecho O, Schumacher J, Marmet Ph, Stenzel O, Büchi FN, Lamibrac A, Münch B (2017) Microstructure-property relationships in a gas diffusion layer (GDL) for polymer electrolyte fuel cells, Part I: effect of compression and anisotropy of dry GDL. Electrochim Acta 227:419–434
Hwang GS, Weber AZ (2012) Effective-diffusivity measurement of partially-saturated fuel-cell gas-diffusion layers. J Electrochem Soc 159:F683–F692
James JP, Choi HW, Pharoah JG (2012) X-ray computed tomography reconstruction and analysis of polymer electrolyte membrane fuel cell porous transport layers. Int J Hydrogen Energy 37:18216–18230
Kashkooli AG, Farhad S, Lee DU, Feng K, Litster S, Babu SK, Zhu L, Chen Z (2016) Multiscale modeling of lithium-ion battery electrodes based on nano-scale X-ray computed tomography. J Power Sources 307:496–509
Khakaz-Babolia M, Harvey DA, Pharoah JG (2012) Investigating the performance of catalyst layer micro-structures with different platinum loadings. ECS Trans 50:765–772
Liu J, García-Salaberri PA, Zenyuk IV (2019) The impact of reaction on the effective properties of multiscale catalytic porous media: a case of polymer electrolyte fuel cells. Transp Porous Med 128:363–384
Mathias M, Roth J, Fleming J, Lehnert W (2003) Diffusion media materials and characterisation. Fuel Cell Technol Appl 3:517
Nabovati A, Llewellin EW, Sousa ACM (2009) A general model for the permeability of fibrous porous media based on fluid flow simulations using the lattice Boltzmann method. Compos Part A 40:860–869
Pant LM, Sushanta KM, Secanell M (2014) Stochastic reconstruction using multiple correlation functions with different-phase-neighbor-based pixel selection. Phys Rev E 90:023306
Pharoah JG, Choi HW, Chueh CC, Harvey DB (2011) Effective transport properties accounting for electrochemical reactions of proton-exchange membrane fuel cell catalyst layers. ECS Trans 41:221–227
Sabharwal M, Pant LM, Putz A, Susac D, Jankovic J, Secanell M (2016) Analysis of catalyst layer microstructures: from imaging to performance. Fuel Cells 16:734–753
Tjaden B, Cooper SJ, Brett DJL, Kramer D, Shearing PR (2016) On the origin and application of the Bruggeman correlation for analysing transport phenomena in electrochemical systems. Curr Opin Chem Eng 12:44–51
Tomadakis MM, Robertson TJ (2005) Viscous permeability of random fiber structures: comparison of electrical and diffusional estimates with experimental and analytical results. J Compos Mater 39:163–188
Tomadakis MM, Sotirchos SV (1993) Ordinary and transition regime diffusion in random fiber structures. AIChE J 39:397–412
Tranter TG, Gostick JT, Burns AD, Gale WF (2018) Capillary hysteresis in neutrally wettable fibrous media: a pore network study of a fuel cell electrode. Transp Porous Med 121:597–620
Wang CY (2004) Fundamental models for fuel cell engineering. Chem Rev 104:4727–4765
Weber AZ, Borup RL, Darling RM, Das PK, Dursch TJ, Gu W, Harvey D, Kusoglu A, Litster S, Mench MM, Mukundan R, Owejan JP, Pharoah JG, Secanell M, Zenyuk IV (2014) A critical review of modeling transport phenomena in polymer-electrolyte fuel cells. J Electrochem Soc 161:F1254–F1299
Zenyuk IV, Parkinson DY, Liam GC, Weber AZ (2016) Gas-diffusion-layer structural properties under compression via X-ray tomography. J Power Sources 328:364–376
Zhang D, Forner-Cuenca A, Oluwadamilola OT, Yufit V, Brushett FR, Brandon NP, Gu S, Cai Q (2020) Understanding the role of the porous electrode microstructure in redox flow battery performance using an experimentally validated 3D pore-scale lattice Boltzmann mode. J Power Sources 447:227249

# Modeling Vanadium Redox Flow Batteries Using OpenFOAM

**Sangwon Kim, Dong Hyup Jeon, Sang Jun Yoon, and Dong Kyu Kim**

**Abstract** This chapter establishes that OpenFOAM is applicable for analyzing the electrolyte flow in a vanadium redox flow battery (VFB) and the transport phenomena in these systems. The local porosity was controlled by inserting an extra layer of electrode at the inlet and outlet. The variations in electrochemical characteristics and energy conversion efficiency with porosity were obtained through VFB single cell experiments. Numerical analysis of the electrolyte flow and pressure distribution provided a theoretical explanation for the physical phenomenon, which depending on the local porosities. When the current density was 50 mA/cm$^2$, the electrode with uniform porosity (UP) showed the best energy performance. However, at a high current density of 150 mA/cm$^2$, the partial porosity-lowered electrode at inlet (designated as LPI) showed better efficiency than the UP electrode since the rate of electrochemical reaction increases, and the mass transfer of the reactant is enhanced accordingly. OpenFOAM is expected to contribute significantly to the optimization

The original version of this chapter was revised: The ESM material has been updated. The correction to this chapter is available at https://doi.org/10.1007/978-3-030-92178-1_10

**Supplementary Information** The online version contains supplementary material available at https://doi.org/10.1007/978-3-030-92178-1_5.

S. Kim (✉)
Korea Institute of Science and Technology Europe, Saarbrücken, Germany
e-mail: sangwon.kim@kist-europe.de

Transfercenter Sustainable Electrochemistry, Saarland University, Saarbrücken, Germany

D. H. Jeon
Department of Mechanical System Engineering, Dongguk University, Gyeongju, Republic of Korea
e-mail: jeondh@dongguk.ac.kr

S. J. Yoon
Advanced Material Division, Korea Research Institute of Chemical Technology, Daejeon, Republic of Korea
e-mail: sjyoon@krict.re.kr

D. K. Kim
School of Mechanical Engineering, Chung-Ang University, Seoul, Republic of Korea
e-mail: dkyukim@cau.ac.kr

© Springer Nature Switzerland AG 2022, corrected publication 2022 
S. Beale and W. Lehnert (eds.), *Electrochemical Cell Calculations with OpenFOAM*, Lecture Notes in Energy 42, https://doi.org/10.1007/978-3-030-92178-1_5

of this flow battery system. In the near future, it will also aid in achieving carbon neutrality by the virtue of collective intelligence.

**Keywords** All-vanadium redox flow battery · Carbon felt electrode · Porous medium flow · Local porosity · High current density

## Nomenclature

| | |
|---|---|
| UP | uniform porosity |
| LPI | low porosity at the inlet |
| LPO | low porosity at the outlet |
| $E_{eq}$ | equilibrium cell potential |
| $E^0$ | standard cell potential |
| $E^{0'}$ | formal cell potential |
| $C_{\mathrm{i}}$ | concentration of the *i* species |
| $C_i^*$ | bulk concentration of the *i* species |
| $R$ | ideal gas constant, 8.314 J/mol K |
| $T$ | cell temperature, K |
| $F$ | Faraday constant, 96,485 A s/mol |
| $k^0$ | standard rate constant |
| $i$ | current at the electrodes |
| CE | current efficiency, Coulomb efficiency |
| VE | voltage efficiency |
| EE | energy efficiency |
| $I$ | current |
| $V$ | voltage |
| $t$ | time |
| $\vec{u}$ | flow velocity vector, m/s |
| $p$ | pressure, N/m$^2$ |
| $\vec{f}$ | acceleration vector caused by body force, m/s$^2$ |
| $S$ | sink term of the momentum loss per unit volume, kg/m$^2$s$^2$ |
| $K$ | permeability of the porous medium, m$^2$ |
| $V_f$ | void volume occupied by fluid phase |
| $V_s$ | void volume occupied by solid phase |
| $V$ | local representative elementary volume |
| REV | representative elementary volume |
| <> | superficial average or phase average (macroscopic) |
| $\langle\rangle^k$ | intrinsic phase average (microscopic) |
| $u_D$ | Darcy velocity (flux), filtration (filter) velocity, superficial velocity |
| $u_p$ | physical velocity, pore velocity, interstitial velocity |
| $D_p$ | mean pore diameter, m |

Fo Forchheimer number
$C_f$ Forchhemier coefficient, inertia coefficient
Re Reynolds number

## Greek

$\alpha$ transfer coefficient, symmetry factor, in Eqs. (8) and (9)
$\alpha$ viscous resistance coefficient, $m^{-2}$, in Eqs. (29), (30), and (32)
$\eta$ overpotential
$\rho$ density, $kg/m^3$
$\rho_0$ reference density or surface density, $kg/m^3$
$\varepsilon$ porosity
$\tau_{ij}$ shear stress tensor, $N/m^2$
$\mu$ dynamic viscosity, Pa·s
$\mu_e$ effective viscosity, Pa·s
$\beta$ inertial resistance coefficient, Forchhemier or non-Darcy coefficient, $m^{-1}$
$\psi_k$ local microscopic property with *k*-phase

## Superscripts

0 standard state when activity is 1.0
' formal state
* bulk solution

## Subscripts

*dis* discharge
*ch* charge
+ positive electrode
− negative electrode
*f* fluid phase
*s* solid phase
*k* phase, solid or fluid
*D* Darcy
*p* physical value or pore
e effective value
0 reference or surface

# 1 Introduction

The entire world is affected by the climate change and global warming caused due to greenhouse gases (Dunn et al. 2011; Armand and Tarascon 2008; Yang et al. 2011). Many efforts are being made worldwide to convert energy sources from fossil fuels, the main cause of greenhouse gas emissions, to renewable energy (Halls et al. 2013). Renewable energies, such as solar and wind energies, are affected by weather and season, and therefore reliable energy storage devices with large capacities are required to ensure stable energy supply to the grid from renewable energy sources (Yang et al. 2011; Larcher and Tarascon 2014).

Lithium ion batteries (LIB) are considered as the best option for energy storage since they have high energy densities (Blomgren 2016) and are one of the most commercialized batteries (Duh et al. 2018). However, LIBs often experience thermal runaway failure, which occurs due to self-sustaining exothermic reactions and an uncontrolled increase in the temperature, resulting in the risk of fire and even explosions (Bravo Diaz et al. 2020; Chen et al. 2021; Kim et al. 2007). The structure of redox flow batteries (RFBs) is completely different from those of LIBs or conventional batteries. This is because the energy carriers are not a part of the electrodes within the battery containment but are contained in two separate external liquid reservoirs (Soloveichik 2015; Ye et al. 2018). This ascertains that there is no risk of fire in an RFB system. The advantage of RFBs over conventional batteries is that the energy storage capacity (in kWh) and power (in kW) can be controlled independently by manipulating the size of the tanks and the surface area/catalytic activity of the electrodes, respectively (Zhang et al. 2017). Owing to this, RFBs are considered more appropriate than LIBs as a large-scale energy storage system (ESS).

There are several RFB technologies including polysulfide/bromide, all-vanadium, and Fe/Cr technologies (Noack et al. 2015; Chen et al. 2017). The all-vanadium redox flow battery (VFB) has received significant attention because of its excellent electrochemical reversibility, high roundtrip efficiency, and negligible cross-contamination between the catholyte and anolyte (Rychcik and Skyllas-Kazacos 1988; Kim 2019; Kim et al. 2020). The major disadvantage of VFBs is that the energy density is much lower than that of LIBs (Cong et al. 2013). To overcome this drawback and to optimize the large-scale energy storage system, ample efforts, including conducting research based on numerical analyses and experimentations, will be required.

Numerical studies have been conducted on VFBs using commercial softwares such as ANSYS FLUENT (Park et al. 2020a, b), STAR-CCM + (Emmel et al. 2020; Prumbohm and Wehinger 2019), and COMSOL (Bromberger et al. 2014). Although commercial softwares provide computation results conveniently and are easy to use, they require high license fees and moreover, the source code cannot be obtained. On the other hand, OpenFOAM (Open Field Operation and Manipulation) is a development framework for computational fluid dynamics (CFD) based on General Public License (GPL) and an open source platform aiming to create better programs with collective intelligence through an open environment like Linux (Weller et al. 1998).

Though most commercial softwares are developed for general purposes, an experienced user can develop exclusive codes for specific problems in a relatively short time using OpenFOAM by applying an independent physical model (Lohaus et al. 2019).

The purpose of this chapter is to establish that OpenFOAM can be applied to numerical analyses in the field of flow batteries. This chapter will also encourage readers to develop their own OpenFOAM code to solve problems of interest. The electrolyte flow has the characteristics of being able to flow through a porous medium because the flow battery electrode is composed of carbon felt. Porosity is the main factor influencing the electrolyte flow and mass transport of the active species. Higher the current density, faster will be the rate of the electrochemical reaction. To improve the energy conversion efficiency of flow batteries, it is necessary to maintain a balance between rate of reactant mass transport and the rate of the redox reaction. If the porosity can be reduced only in the efficient area instead of the entire active area to balance the two rates, the carbon felt can be reduced. This would further ensure that the electrolyte flows well, with less resistance.

Based on the above viewpoint, we considered three different types of electrode porosity: uniform porosity (UP), low porosity at the inlet (LPI), and low porosity at the outlet (LPO).

The first part of this chapter explains the fundamentals of vanadium redox batteries and the flow characteristics in a porous medium. Subsequently, the experimental configuration and modeling implementation are described. Next, the experimental and simulation results are compared and discussed. Finally, conclusions are drawn, and the future prospects are discussed.

## 2 Overview of Vanadium Redox Flow Batteries

The structure of a redox flow battery similar to that of a polymer electrolyte membrane fuel cell in a stack configuration (Fig. 1). The redox flow battery deals only with the single-phase flow of the electrolyte, while the PEM fuel cell involves the two-phase flow of gas and liquid. The redox flow battery charges and discharges electric energy according to the change in the oxidation number of the active material of the electrolyte based on the principles of redox reactions.

VFBs use vanadium ions of different oxidation numbers as the active materials for electrolytes. When trivalent ($V^{3+}$) and tetravalent ($V^{4+}$) vanadium are added to the negative and positive electrodes, respectively, and electricity is applied to each electrode during the charging process, trivalent vanadium ($V^{3+}$) is reduced to divalent vanadium ($V^{2+}$) at the negative electrode, and tetravalent vanadium ($V^{4+}$) is oxidized to pentavalent vanadium ($V^{5+}$) at the positive electrode. During this, the difference between the oxidation numbers of the anode and cathode increases from 1 to 3, and it can be conceptually explained that this difference in the oxidation number is related to the storage of the converted electrochemical energy. The opposite reaction occurs during discharging (Fig. 2). The chemical reactions for the charge–discharge process

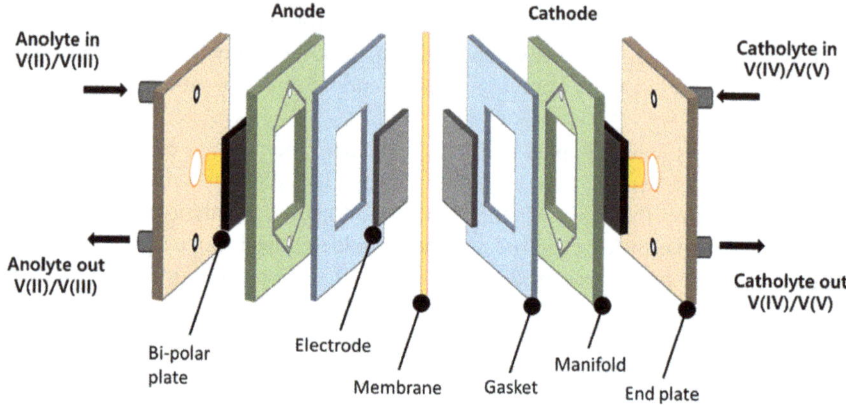


**Fig. 1** Schematic of the configuration of a vanadium redox flow battery single cell assembly. Reproduced with permission from Yoon et al. (2019). Copyright 2019 by Elsevier

are expressed as follows:

Negative electrode:

$$V^{2+} \leftrightarrow V^{3+} + e^{-} E0 = -0.255V \tag{1}$$

Positive electrode:

$$VO_2^+ + e^- + 2H^+ \leftrightarrow VO^{2+} + H_2OE0 = +1.004V \tag{2}$$

Overall reaction:

$$VO_2^+ + V^{2+} + 2H^+ \leftrightarrow VO^{2+} + V^{3+} + H_2OE0 = +1.259V \tag{3}$$

The equilibrium cell potentials, $E_{eq}$, for each reaction are calculated using the Nernst equation (Allen and Bard 2001; John Newman2021) as follows:

$$E_{eq,-} = E_-^0 + \frac{RT}{F} ln\left(\frac{C_{V^{3+}}}{C_{V^{2+}}}\right) \tag{4}$$

$$E_{eq,+} = E_+^0 + \frac{RT}{F} ln\left(\frac{C_{VO_2^+}(c_{H^+})^2}{C_{VO^{2+}}}\right) \tag{5}$$

$$E_{eq,overall} = E_{overall}^0 + \frac{RT}{F} ln\left(\frac{C_{VO_2^+}(C_{H^+})^2}{C_{VO^{2+}}}\frac{C_{V^{2+}}}{C_{V^{3+}}}\right) \tag{6}$$

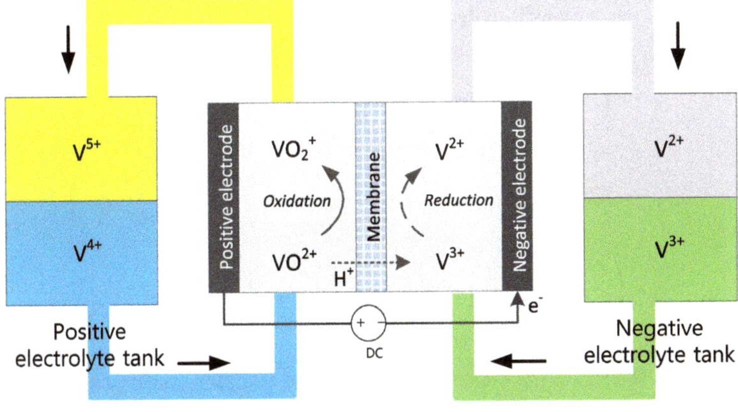


(a) Charging

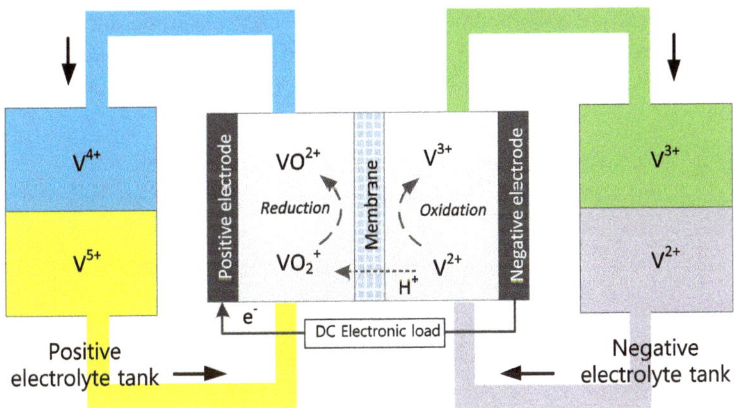


(b) Discharging

**Fig. 2** Operating principle of a vanadium redox flow battery

Here, $C_i$ is the concentration of the $i$ species, $E^0$ is the standard cell potential for the electrode reaction,[1] $R$ is the ideal gas constant (8.314 J/mol K), $T$ is the cell temperature, and $F$ is the Faraday constant (96,485 A s/mol).

The currents at each electrode can be expressed using the Butler-Volmer equation (Allen and Bard 2001; John Newman 2021) as

[1] In order to use concentration instead of activity in the Nernst equation, formal cell potential, $E^{0\prime}$, is a more correct expression than standard cell potential, $E^0$. Since the difference between the two values is usually not significant, $E^{0\prime}$ and $E^0$ are not distinguished in this chapter. (Allen and Bard 2001).

$$i_{-} = Fk_{-}^{0}C_{V^{3+}}^{*(1-\alpha_{-})}C_{V^{2+}}^{*\alpha_{-}}\left[\left(\frac{C_{V^{3+}}(0,t)}{C_{V^{3+}}^{*}}\right)\exp\left(-\frac{\alpha_{-}F}{RT}\eta_{-}\right) - \left(\frac{C_{V^{2+}}(0,t)}{C_{V^{2+}}^{*}}\right)\exp\left(\frac{(1-\alpha_{-})F}{RT}\eta_{-}\right)\right] \quad (7)$$

$$i_{+} = Fk_{+}^{0}C_{VO_2^{+}}^{*(1-\alpha_{+})}C_{VO^{2+}}^{*\alpha_{+}}\left[\left(\frac{C_{VO_2^{+}}(0,t)}{C_{VO_2^{+}}^{*}}\right)\exp\left(-\frac{\alpha_{+}F}{RT}\eta_{+}\right) - \left(\frac{C_{VO^{2+}}(0,t)}{C_{VO^{2+}}^{*}}\right)\exp\left(\frac{(1-\alpha_{+})F}{RT}\eta_{+}\right)\right] \quad (8)$$

Here, $k^0$ is the standard rate constant, $\alpha$ is the transfer coefficient or symmetry factor, and $\eta$ is the overpotential. The relationship between the voltage and current during the charge–discharge reaction is shown in Fig. 3.

The electrochemical reactions take place on the solid surfaces of the porous carbon felt electrode. The vanadium ions diffuse from the bulk electrolyte to the vicinity of the electrode and are absorbed on the surface of each electrode during the charge process as shown in Fig. 4a and b. The absorbed vanadium ions are linked to the electrode via exchange with the hydrogen ion in the C–OH functional group. The

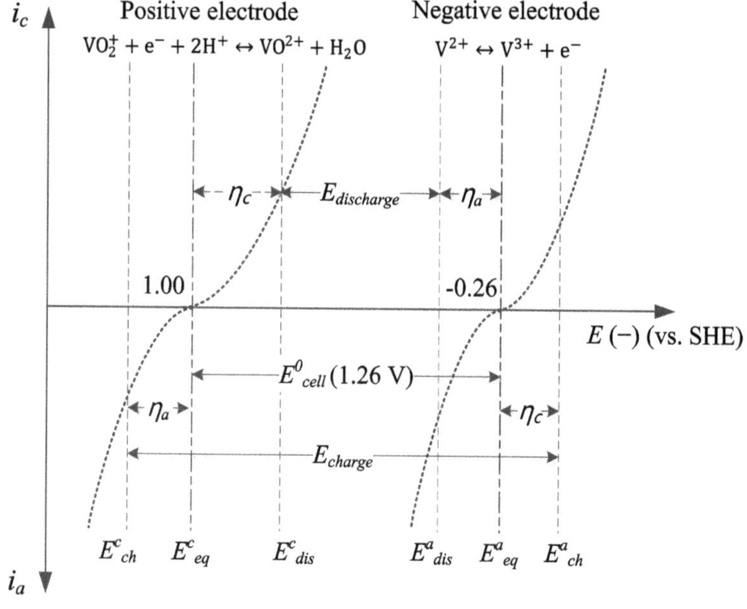


**Fig. 3** Charge–discharge voltage of a vanadium redox flow battery: current vs. voltage, overpotential, and open circuit voltage at the positive and negative electrodes. Reproduced with permission from Kim (2019). Copyright 2019 by Kim S

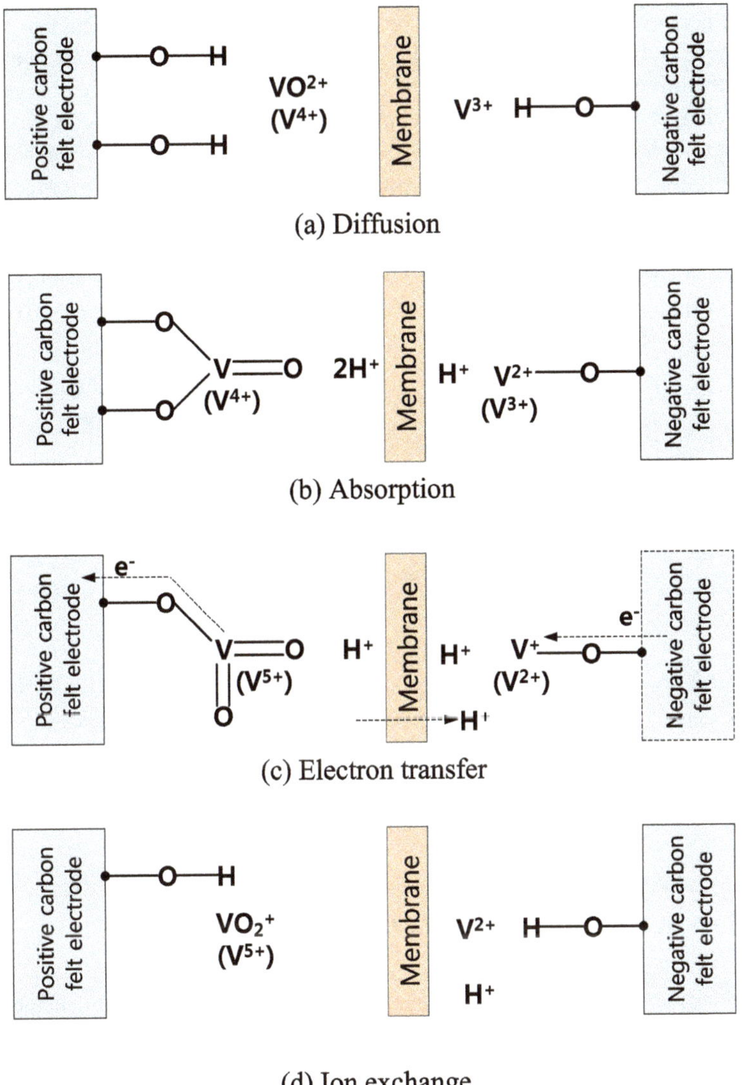


**Fig. 4** Schematic illustration of the redox reactions at the carbon felt electrode surfaces during charging

electron transfer between the linked vanadium ions and the electrode and the accompanying redox reactions are shown in Fig. 4c. Finally, there is ion exchange between the vanadium ions attached to the electrode surface and the hydrogen ion in the electrolyte, and the generated reactants ($VO_2^+$ and $V^{2+}$) diffuse respectively back into the original electrolytes (Kim et al. 2015).

The performance of a VFB can be measured with current efficiency, voltage efficiency, and energy efficiency. The current efficiency (CE, also known as the coulombic efficiency) is defined as the ratio of the amounts of usable charge to the stored charge; that is, the discharge capacity divided by the charge capacity. CE is a measure of the storage capacity loss during the charge–discharge process. Voltage efficiency (VE) is the ratio of the average discharge voltage to the average charge voltage. Voltage efficiency is a measure of the loss in electrical resistance and that of the polarization properties of the battery. Energy efficiency is the ratio of the available energy to the stored energy and can be calculated as the product of the voltage efficiency and current efficiency (Ye et al. 2018; Chen et al. 2017; Kim 2019).

$$\begin{aligned} CE &= \frac{dischargecapacity}{chargecapacity} \times 100\% = \frac{\int_{dis} I(t)dt}{\int_{ch} I(t)dt} \times 100\% \\ &= \frac{I_{dis} \bullet t_{dis}}{I_{ch} \bullet t_{ch}} \times 100\% = \frac{t_{dis}}{t_{ch}} \times 100\% (\text{If}\, I_{dis} = I_{ch}) \end{aligned} \tag{9}$$

$$VE = \frac{averagedischargevoltage}{averagechargevoltage} \times 100\% = \frac{\int_{dis} V(t)dt/t_{dis}}{\int_{ch} V(t)dt/t_{ch}} \times 100\% \tag{10}$$

$$\begin{aligned} EE &= \frac{dischargeenergy(Wh)}{chargeenergy(Wh)} \times 100\% = \frac{\int_{dis} I(t)V(t)dt}{\int_{ch} I(t)V(t)dt} \times 100\% \\ &= \frac{I_{dis} \bullet t_{dis}}{I_{ch} \bullet t_{ch}} \frac{\int_{dis} V(t)dt/t_{dis}}{\int_{ch} V(t)dt/t_{ch}} \times 100\% = CE \times VE \end{aligned} \tag{11}$$

## 3 Flow in a Porous Medium, Carbon Felt Electrode

The electrode of a redox flow battery does not participate directly in the redox reaction but provides an active site for the reaction. Carbon felt is extensively used as an electrode material for VFB because of its large reactive specific surface area, excellent chemical stability to sulfuric acid-based electrolytes, and high electrical conductivity (Kim et al. 2015, 2014). The electrolyte flow in a carbon felt electrode can be regarded as the flow through a porous medium (Yoon et al. 2019).

Transfer of the reactants of the redox active species results in an electrochemical reaction while the electrolyte flows through the porous carbon felt medium. The amount of vanadium ions changes during the charging and discharging processes; however, there is no change in the total volume of the electrolyte during the electrochemical reaction. Therefore, the momentum governing equation for the electrolyte can be modeled by adding the sink term due to the porous media to the Navier-Stoke equation (Yoon et al. 2019; Knudsen et al. 2015; Beale et al. 2016). The sink term shown in Eq. (13) makes the mesh finer, requiring more computational power.

Therefore, it is necessary to simplify the sink term according to the flow conditions. Simplification of the sink term is described later in this section.

$$\rho \frac{D\vec{u}}{Dt} = -\nabla p + \rho \vec{f} + \nabla \bullet \tau_{ij} + S \tag{12}$$

$$S = -\frac{\mu}{K}\vec{u} + \mu \nabla^2 \vec{u} - \beta \rho \left| \vec{u} \right| \vec{u} \tag{13}$$

Here, $\vec{u}$ is the flow velocity vector [m/s], p is the pressure [N/m$^2$], $\vec{f}$ is the acceleration vector caused by body force [m/s$^2$], $\tau_{ij}$ is the shear stress tensor [N/m$^2$], S denotes the sink term of the momentum loss per unit volume caused by the porous medium [kg/m$^2$s$^2$], ρ is the fluid density [kg/m$^3$], $\mu$ is the dynamic viscosity [Pa·s], K is the permeability of the porous medium [m$^2$], and $\beta$ is the Forchheimer or non-Darcy coefficient [m$^{-1}$].

A schematic of the flow in a porous medium is given in Fig. 5. The porous medium is composed of pores (voids) and solid matrix. The smallest differential volume that maintains the statistically meaningful local average properties such as

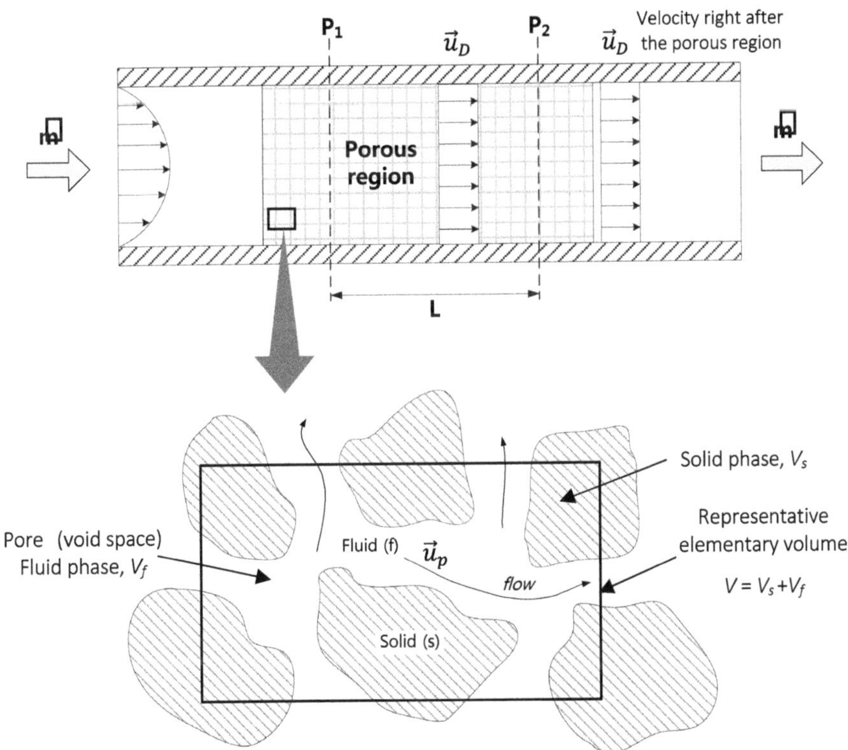


**Fig. 5** Schematic of the flow in a porous medium and the representative elementary volume

porosity, saturation, and pressure is called the representative elementary volume (REV). Porosity is defined as the ratio of the volume fraction occupied by the pores to the total volume; i.e., the quotient of the void volume and the total volume.

$$\varepsilon = \frac{\text{Voidvolume}}{\text{Totalvolume}} = \frac{V_f}{V} = 1 - \frac{V_s}{V} \tag{14}$$

Here, $\varepsilon$ is the porosity, $V_f$ is the void volume occupied by fluid phase, $V_s$ is the volume occupied by solid phase, and $V$ is the local REV. Physically, porosity is the portion through which a fluid can pass in a porous medium and has a value between 0 and 1. Therefore, when porosity is 1, the medium is completely empty, and the fluid can flow freely without any interference.

The volume average quantities are defined depending on the assigned averaging volume (Hsu and Cheng 1990; Whitaker 1996). The quantity associated with the $k$-phase[2] is $\psi_{k,}$, which is basically denotes the local microscopic properties. The superficial average or phase average (macroscopic) is defined as

$$\langle \psi_k \rangle = \frac{1}{V} \int_{V_k} \psi_k dV \tag{15}$$

The intrinsic phase average (local macroscopic) is defined as

$$\langle \psi_k \rangle^k = \frac{1}{V_k} \int_{V_k} \psi_k dV \tag{16}$$

The relationship between these two averages is given by Eq. 17,

$$\langle \psi_k \rangle = \frac{V_k}{V} \frac{1}{V_k} \int_{V_k} \psi_k dV v = \varepsilon_k \langle \psi_k \rangle^k \tag{17}$$

where $\varepsilon_k = V_k/V$ is the local volume fraction of the $k$-phase.

The flow discharge rate in the porous medium is proportional to the pressure drop over a given distance and permeability, K [m$^2$], and is inversely proportional to the fluid viscosity. This equation is well known as Darcy's law or Darcy's equation.

$$\vec{u}_D = -\frac{K}{\mu} \frac{(P_2 - P_1)}{L} \approx -\frac{K}{\mu} \frac{dp}{dx} \approx -\frac{K}{\mu} \nabla p \tag{18}$$

The negative sign for the pressure gradient indicates that the fluid flows in the direction of decreasing pressure. The flow discharging rate, $u_D$, is known as the Darcy velocity (flux) or filtration (filter) velocity or superficial velocity.

The physical velocity in the pore, $u_p$, is higher than the Darcy velocity by a factor of $1/\varepsilon$, since the Darcy velocity is the superficial average value based on the entire

[2] $k \in \{f, s\}$, $f$ and $s$ represent the fluid phase and solid phase, respectively.

volume while the physical velocity is the corresponding intrinsic phase averaged value (Dupuit-Forchheimer relation) (Kaviani 1995). The physical velocity is also called the pore velocity or interstitial velocity.

$$\vec{u}_p \geq \vec{u}_D, \vec{u}_p = \frac{\vec{u}_D}{\varepsilon}, 0 \leq \varepsilon \leq 1 \tag{19}$$

Darcy's law is valid only when the flow speed is low enough that the velocity head is negligible and the viscous force is dominant over the inertial force with a Reynolds number less than one. The Darcy equation can be rewritten for the pressure term as follows:

$$\nabla p = -\frac{\mu}{K}\vec{u}_D \tag{20}$$

Darcy's law can be compared with the viscous penetration-dominated flow, which is also the Stokes flow (Kaviani 1995).

$$\nabla p = \frac{\mu}{\varepsilon}\nabla^2\vec{u}_D = \mu\nabla^2\vec{u}_p \tag{21}$$

Brinkman (Brinkman 1949; Liu et al. 1994) superimposed the diffusion term of the Stokes flow in the Darcy flow as follows.

$$\nabla p = -\frac{\mu}{K}\vec{u}_D + \mu_e\nabla^2\vec{u}_D \tag{22}$$

Here, $\mu_e$ is the effective viscosity and is experimentally similar to the dynamic viscosity of the fluid.

The Reynolds number using the square root of the permeability or the mean pore diameter as a characteristic length was proposed to clarify the flow regimes.

$$Re_K = \frac{\rho u_D\sqrt{K}}{\mu} \tag{23}$$

$$Re_{D_p} = \frac{\rho u D_p}{\mu} \tag{24}$$

Here, $K$ is the permeability of the porous medium [m$^2$], and $D_p$ is the mean pore diameter [m].

When $Re_K$ is less than 1, the porous flow is in the Darcy regime or creeping flow regime. When $0 < Re_K < 10$, the flow is dominated by the inertial effects rather than the viscous effects, causing non-linearity of the governing equation. The Forchheimer regime, where the inertia force is superior to the viscous force, is governed by the following Forchheimer equation (Whitaker 1996; Amiri et al. 2019; Mahdi et al. 2015).

$$\nabla p = -\left(\frac{\mu}{K}\vec{u}_D + \beta\rho\left|\vec{u}_D\right|\vec{u}_D\right) = -\frac{\mu}{K}\vec{u}_D\left(1 + \frac{\beta K\rho\left|\vec{u}_D\right|}{\mu}\right)$$
$$= -\frac{\mu}{K}\vec{u}_D(1 + Fo) \tag{25}$$

Here, $\beta$ is the inertial resistance coefficient or the non-Darcy coefficient [1/m]. The velocity square term, $\beta\rho\left|\vec{u}_D\right|\vec{u}_D$, is called the Forchheimer term while the linear term of the velocity, $\frac{\mu}{K}\vec{u}_D$, is called the Darcy term.

The Forchheimer number (Amiri et al. 2019) is defined as

$$Fo = \frac{\beta K\rho\left|\vec{u}_D\right|}{\mu} \tag{26}$$

The inertial resistance coefficient is defined as

$$\beta = \frac{C_f}{\sqrt{K}} \tag{27}$$

Here, $C_f$ is the Forchheimer coefficient or inertia coefficient.

Forchheimer number $Fo$ and Forchheimer coefficient $C_f$ are related as follows:

$$Fo = \frac{\beta K\rho\left|\vec{u}_D\right|}{\mu} = \beta\sqrt{K} \bullet \frac{\rho U\sqrt{K}}{\mu} = C_f \bullet Re_K \tag{28}$$

The viscous resistance coefficient, $\alpha$, is defined as the inverse of the permeability.

$$\alpha = \frac{1}{K} \tag{29}$$

Therefore, Forchheimer equation can be rewritten as

$$-\nabla p = \alpha\mu\vec{u}_D + \beta\rho\left|\vec{u}_D\right|\vec{u}_D = \frac{\mu}{K}\vec{u}_D + \frac{\rho C_f}{\sqrt{K}}\left|\vec{u}_D\right|\vec{u}_D \tag{30}$$

Thus, by combining all the above equations, the governing momentum equation for the flow through a porous medium can be expressed as follows (Kaviani 1995; Nield and Bejan 2006; Vafai and Kim 1995):

$$\frac{\rho_0}{\varepsilon}\left(\frac{\partial\vec{u}_D}{\partial t} + \frac{1}{\varepsilon}\vec{u}_D \bullet \nabla\vec{u}_D\right) = -\nabla\langle p\rangle^f + \rho\vec{f} + \frac{\mu}{\varepsilon}\nabla^2\vec{u}_D$$
$$\frac{\mu}{K}\vec{u}_D - \frac{\rho C_f}{\sqrt{K}}\left|\vec{u}_D\right|\vec{u}_D \tag{31}$$

# 4 Methods

The purpose of this study is to increase the current density by controlling the local porosity of the electrode so that the electrolyte is evenly distributed over the reaction area of the carbon felt electrode. The effect of porosity on the charge/discharge characteristics and energy conversion efficiency was investigated experimentally, and the flow characteristics were studied using numerical analysis. The schemes for experiments and numerical analysis are described in the following sections.

## 4.1 Experiment

A VFB single cell with an active area of 5 cm × 5 cm manufactured by CNL Energy was used in this study. The positive and negative electrode chambers were constructed of PVC frame, and a Nafion® 115 membrane was sandwiched by the carbon felt separating the chambers. Carbon felt (SGL group, GFD4.6 EA) was used as the electrode material to provide reaction sites and a channel for the electrolyte. Graphite bi-polar plate and copper current collector were connected to the carbon felt. The electrolyte contained 1.6 M $V^{3+}$ and 1.6 M $V^{4+}$ dissolved in 3 M $H_2SO_4$ (30 mL). The experimental setup is shown in Fig. 6. The electrolyte flow rate was 13 mL/min, and the current density was varied from 50 to 150 mA/cm$^2$ in the battery cycling test. The cut-off voltages for charging and discharging were 1.6 and 1.0 V, respectively.

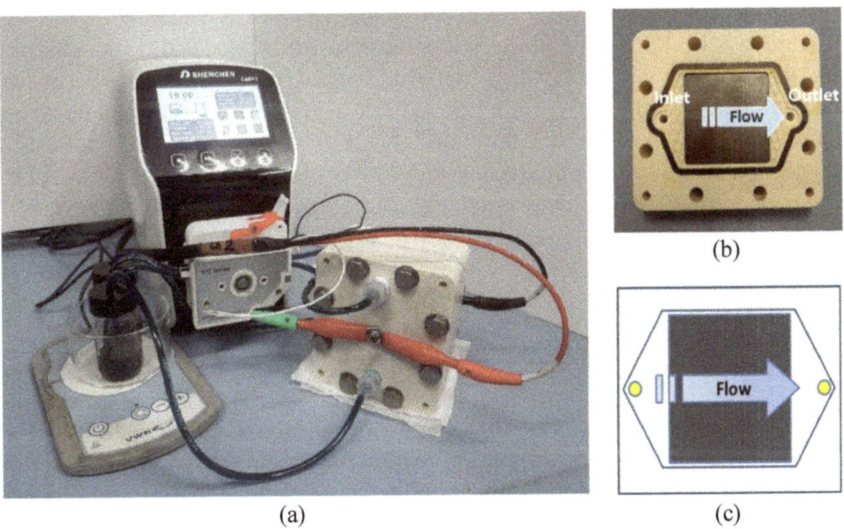


**Fig. 6** Schematic of all-vanadium redox flow battery experiment: **a** experimental set up for VFB, **b** half-cell configuration of VFB, **c** schematic of electrolyte flow direction

Fig. 7 Carbon felt electrodes with different local porosities

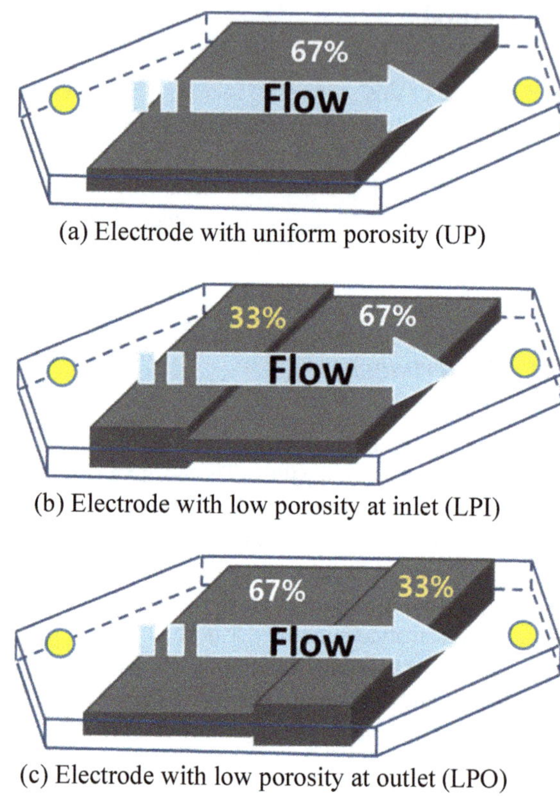


(a) Electrode with uniform porosity (UP)

(b) Electrode with low porosity at inlet (LPI)

(c) Electrode with low porosity at outlet (LPO)

Carbon felt electrodes of three different porosities were considered. The first was the reference electrode with a uniform porosity (UP) of 67%. For the second electrode, the porosity was locally lowered (33%) at the electrolyte inlet by adding carbon felt at the inlet. For the third electrode, the porosity was locally reduced at the electrolyte outlet. The electrodes with low porosity at the inlet and outlet are designated as LPI and LPO, respectively. Three electrode designs are depicted in Fig. 7.

## *4.2 Implementation of Numerical Model*

In this study, we developed a two-dimensional simulation model to analyze the flow field of the electrolyte in the three different porous electrodes. The meshes for the OpenFOAM calculations are shown in Fig. 8.

The program code for solving the Navier–Stokes equation is given below.

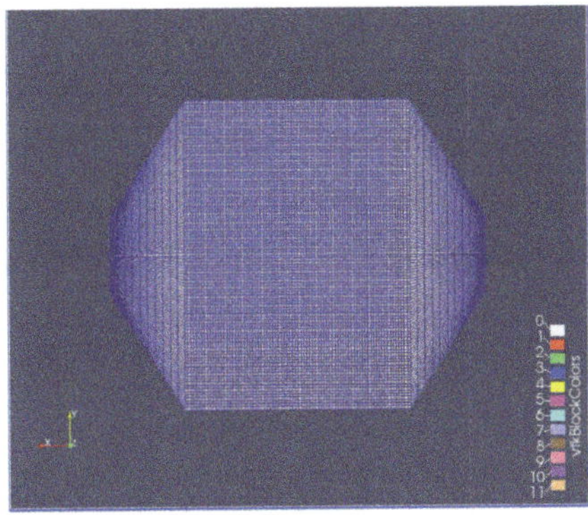
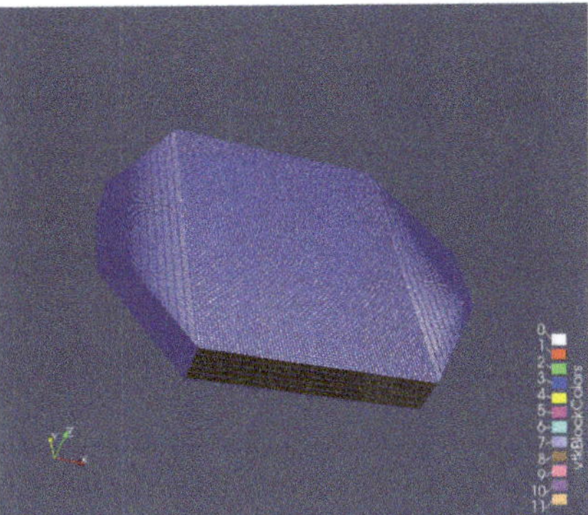

**Fig. 8** VFB half-cell mesh

```
fvVectorMatrix UEqn
    (fvm::div(phiCath, UCath) - fvm::laplacian(nuCath, UCath));
    UEqn.relax();
    solve(UEqn == -fvc::grad(pCath));
    for (corr = 0; corr < nCorr; corr++)
    {
        pCath.boundaryField().updateCoeffs();
        UCath = UEqn.H()/UEqn.A();
        UCath.correctBoundaryConditions();
        phiCathl = fvc::interpolate(UCath) & Cath-Mesh.Sf();
        for (nonOrth = 0; nonOrth <= nNonOrthCorr; non-Orth++)
        {
            fvScalarMatrix pEqn
            (fvm::laplacian(1.0/UEqn.A(), pCath) == fvc::div(phiCath));
            pEqn.setReference(pCathRefCell, pCathRefValue);
            pEqn.solve();
            if (nonOrth == nNonOrthCorr) phiCath -= pEqn.flux();
        }
        continuityErrs(phiCath);
pCath.relax();
        UCath -= fvc::grad(pCath)/UEqn.A();
        UCath.correctBoundaryConditions();
    }
}
```

**Table 1** Electrode and electrolyte parameters used for simulation

| Parameters | Value |
|---|---|
| Compressed thickness of electrode | 3 mm |
| Uniform porosity of electrode | 0.67 |
| Locally lowered porosity of electrode | 0.33 |
| Dynamic viscosity of electrolyte | $4.928 \times 10^{-3}$ Pa·s |
| Density of electrolyte | 1.84 g/cm$^3$ |
| Diffusion coefficient of electrolyte | $2.12 \times 10^{-9}$ m$^2$/s |

The Darcy-Forchheimer equation is used as the governing equation for the sink term in this simulation.

$$S_i = \left(\alpha\mu + \frac{1}{2}\rho C_f\left|\vec{u}_D\right|\right)\vec{u}_D \tag{32}$$

Here, $\alpha = \frac{1}{K}$ and $C_f = 2\beta$.

All the parameters associated with the simulation are listed in Table 1. In this study, the open-source software, OpenFOAM version 2.1.2, was used for the two-phase flow simulation within a porous electrode.

## 5 Results and Discussion

The electrolyte distribution in the three different electrodes is shown in Fig. 9. The dark blue region represents the electrolyte, and the white spots represent the voids. The electrochemical reaction does not occur in the void, suggesting the loss of active area for the redox reaction. It is obvious that the performance of the VFB will

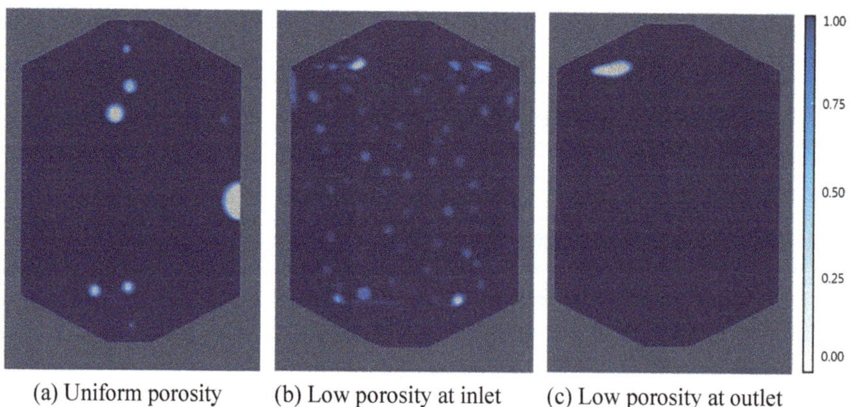


(a) Uniform porosity (b) Low porosity at inlet (c) Low porosity at outlet

**Fig. 9** Electrolyte distribution in the three electrodes

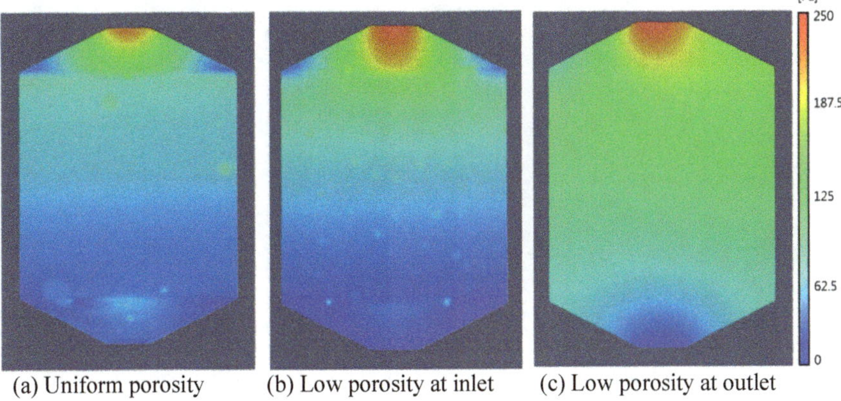


**Fig. 10** Static pressure distribution in the three electrodes

improve as the blue area increases and the white area decreases. In the UP electrode, the electrolyte cannot reach the specific region, because of which there are several medium-sized air bubbles. In the LPI electrode, small air bubbles are observed in some places, and it can be confirmed that the electrolyte evenly contacts the electrode in all the regions. In the LPO electrode, a large empty space is formed in the inlet area since the flow in the outlet is stagnant relative to that in the inlet due to the low porosity at the former. This can be clearly presented in the static pressure contours in Fig. 10. The pressure drop through the porous UP electrode was 0.18 bar, which is relatively low compared to that for the LPI and LPO electrodes, both of which were 0.27 bar. In addition, the pressure drop was linear within the UP electrode, while it was nonlinear for the other two electrodes with non-uniform porosity. Owing to the high tortuosity, a large pressure drop was observed at the point in which the porosity changed.

Analysis of the velocity vectors in Fig. 11 provides further insights into the electrolyte distribution in the electrode. Except for the inlet region, the flow vectors in the UP electrode were neatly aligned without mixing while those in the LPI and LPO electrodes were well mixed due to the local non-uniform porosity. The disordered flow can enhance the mass transport of the reactant and product and increase the contact of the electrolyte with the electrode surface, which would increase the energy efficiency of the VFBs at high current densities. Therefore, although high tortuosity lowers the flow rate, it facilitates the electrochemical reaction by increasing the chances of the electrolyte coming in contact with the electrode surface.

The charge–discharge behavior of the VFBs with different electrodes at a current density of 50 and 125 mA/cm$^2$ and a flow rate of 13 mL/min is shown in Fig. 12. At the beginning of the charging process, the three cells had similar starting voltages. However, the operating voltage reached 1.6 V faster for the cells with the LPI electrode and LPO electrode than for the cell with the UP electrode. Similarly, the discharge speed was faster for the cells with the LPI and LPO electrodes than for the

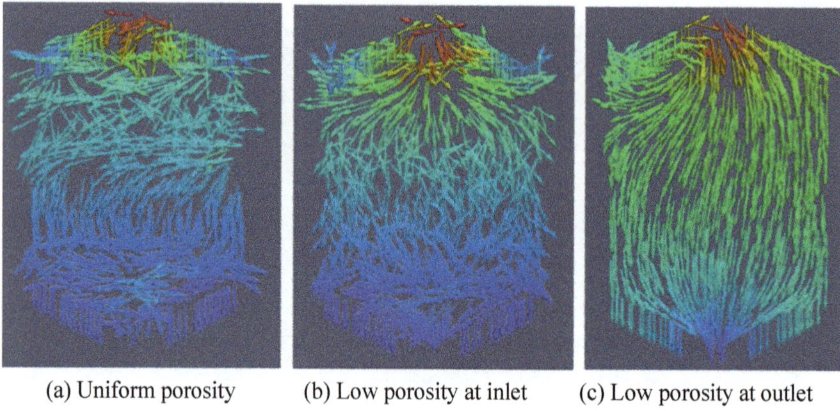

(a) Uniform porosity (b) Low porosity at inlet (c) Low porosity at outlet

**Fig. 11** Velocity vector plots of the three electrodes

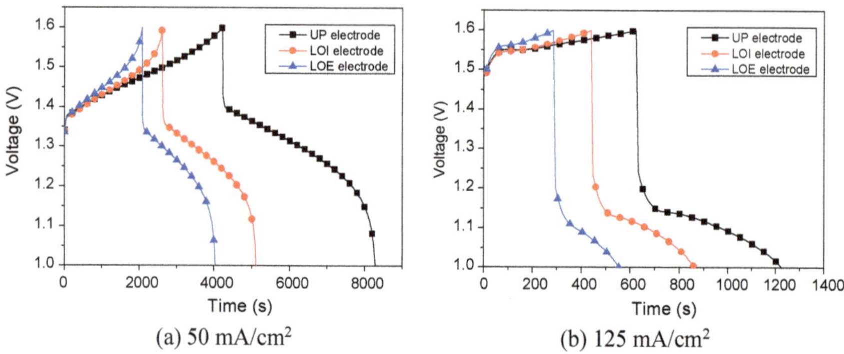


(a) 50 mA/cm$^2$ (b) 125 mA/cm$^2$

**Fig. 12** Charge–discharge curves of VFB with different electrodes at a flow rate of 13 mL/min

cell with the UP electrode. It can be seen that the charge–discharge curve of the LPI electrode approaches that of the UP electrode with increasing current density.

The efficiencies of the three electrodes between 50 and 150 mA/cm$^2$ at a flow rate of 13 mL/min were compared (Fig. 13). At low current densities, the UP electrode showed the best efficiency since the electrolyte was distributed uniformly through the carbon felt due to the uniform porosity in entire active area. However, at higher current densities, efficiency of the LPI electrode was better than that of the UP electrode. As the current density increases, a larger electrode surface area is required to balance the mass transfer for enhancing the rate of the electrochemical reaction. When carbon felt was added locally, the porosity decreased, surface area increased, and conductivity improved locally. When the porosity was partially reduced, it was found that the reduction in porosity on the inlet side was more effective than that on the outlet side of the electrolyte.

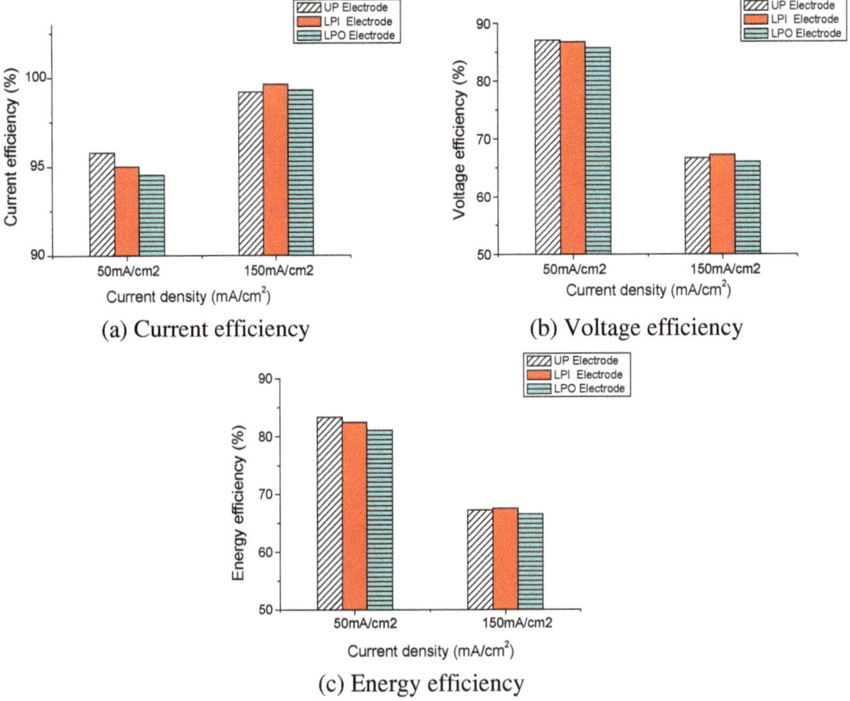


(a) Current efficiency

(b) Voltage efficiency

(c) Energy efficiency

**Fig. 13** Efficiency of the different electrodes between 50 and 150 mA/cm$^2$ at a flow rate of 13 mL/min

## 6 Conclusions

In this chapter, it has been shown that the electrolyte flow of a vanadium redox flow battery (VFB) can be successfully analyzed using OpenFOAM. We investigated the flow through a porous medium of carbon felt electrode; the local porosities were controlled by inserting an extra layer of electrode at the inlet and outlet.

The porosity-dependent charge–discharge characteristics and energy conversion efficiencies were obtained through VFB single cell experiments. The OpenFOAM simulation was conducted to explain the experimental results and to obtain a sound theoretical basis for explaining the physical phenomenon.

At low current densities, the electrode with uniform porosity (UP) showed the best energy efficiency. However, with increasing current density, the rate of electrochemical reaction increased, and an enhanced mass transfer of the reactant was required. The partial porosity-lowered electrode at inlet (LPI) showed better efficiency than the UP electrode. In addition, when the porosity was reduced partially, it was more effective to reduce the porosity at the inlet rather than at the outlet of the electrolyte.

The electrochemical reaction was not included in the current OpenFOAM numerical analysis. The next step is to include the electrochemical reactions in the OpenFOAM calculation in order to investigate the correlation between the rate of the electrochemical reaction and that of the mass transfer at varying current densities. It would be meaningful to confirm the applicability of OpenFOAM to numerical analysis of VFBs, which is one of the most promising candidates as large-scale energy storage devices. OpenFOAM, an open-source software, is expected to contribute increasingly to the optimization of flow battery systems by the virtue of collective intelligence and is expected to significantly contribute in achieving carbon neutrality.

**Acknowledgements** This research was supported by Korea Institute of Science and Technology (2E30993), Dongguk University research fund of 2021, the KRICT Core Research Program (KK2022-20) and Chung Ang University research fund of 2021.

## References

Allen J,.Bard LRF (2001) Electrochemical methods: fundamentals and applications, 2nd edn

Amiri L, Ghoreishi-Madiseh SA, Hassani FP, Sasmito AP (2019) Estimating pressure drop and Ergun/Forchheimer parameters of flow through packed bed of spheres with large particle diameters. Powder Technol 356:310–324. https://doi.org/10.1016/j.powtec.2019.08.029

Armand M, Tarascon JM (2008) Building better batteries. Nature 451:652–657. https://doi.org/10.1038/451652a

Beale SB, Choi HW, Pharoah JG et al (2016) Open-source computational model of a solid oxide fuel cell. Comput Phys Commun 200:15–26. https://doi.org/10.1016/j.cpc.2015.10.007

Blomgren GE (2016) The development and future of lithium ion batteries. J Electrochem Soc

Bravo DL, He X, Hu Z et al (2020) Review—meta-review of fire safety of lithium-ion batteries: industry challenges and research contributions. J Electrochem Soc 167:090559. https://doi.org/10.1149/1945-7111/aba8b9

Brinkman HC (1949) A calculation of the viscous force exerted by a flowing fluid on a dense swarm of particles. Appl Sci Res 1:27–34. https://doi.org/10.1007/BF02120313

Bromberger K, Kaunert J, Smolinka T (2014) A model for all-vanadium redox flow batteries: introducing electrode-compression effects on voltage losses and hydraulics. Energy Technol 2:64–76. https://doi.org/10.1002/ente.201300114

Chen Y, Kang Y, Zhao Y et al (2021) A review of lithium-ion battery safety concerns: the issues, strategies, and testing standards. J Energy Chem 59:83–99. https://doi.org/10.1016/j.jechem.2020.10.017

Chen R, Kim S, Chang Z (2017) Redox flow batteries: fundamentals and applications. Redox—principles and advanced applications. InTech, pp 103–118

Cong D, Huamin Z, Xianfeng L et al. (2013) Vanadium flow battery for energy storage: prospects and challenges. J Phys Chem Lett 4:1281–1294

Larcher D, Tarascon JM (2014) Towards greener and more sustainable batteries for electrical energy storage. Nat Chem

Duh YS, Lin KH, Kao CS (2018) Experimental investigation and visualization on thermal runaway of hard prismatic lithium-ion batteries used in smart phones. J Therm Anal Calorim 132:1677–1692. https://doi.org/10.1007/s10973-018-7077-2

Dunn B, Kamath H, Tarascon JM (2011) Electrical energy storage for the grid: a battery of choices. Science (80)334:928–935. https://doi.org/10.1126/science.1212741

Emmel D, Hofmann JD, Arlt T et al (2020) Understanding the impact of compression on the active area of carbon felt electrodes for redox flow batteries. ACS Appl Energy Mater 3:4384–4393. https://doi.org/10.1021/acsaem.0c00075

Halls JE, Hawthornthwaite A, Hepworth RJ et al (2013) Empowering the smart grid: can redox batteries be matched to renewable energy systems for energy storage? Energy Environ Sci 6:1026–1041. https://doi.org/10.1039/c3ee23708g

Hsu CT, Cheng P (1990) Thermal dispersion in a porous medium. Int J Heat Mass Transf. https://doi.org/10.1016/0017-9310(90)90015-M

Kaviani M (1995) Principles of heat transfer in porous media, 2nd edn 95 (corrected second printing 1999)

Kim GH, Pesaran A, Spotnitz R (2007) A three-dimensional thermal abuse model for lithium-ion cells. J Power Sources 170:476–489. https://doi.org/10.1016/j.jpowsour.2007.04.018

Kim KJ, Park MS, Kim YJ et al (2015) A technology review of electrodes and reaction mechanisms in vanadium redox flow batteries. J Mater Chem A. https://doi.org/10.1039/c5ta02613j

Kim S (2019) Vanadium redox flow batteries: electrochemical engineering. Energy storage devices. InTech, pp 115–133

Kim KJ, Lee SW, Yim T et al (2014) A new strategy for integrating abundant oxygen functional groups into carbon felt electrode for vanadium redox flow batteries. Sci Rep 4. https://doi.org/10.1038/srep06906

Kim DK, Yoon SJ, Kim S (2020) Transport phenomena associated with capacity loss of all-vanadium redox flow battery. Int J Heat Mass Transf 148: 119040. https://doi.org/10.1016/j.ijheatmasstransfer.2019.

Knudsen E, Albertus P, Cho KT et al (2015) Flow simulation and analysis of high-power flow batteries. J Power Sour 299:617–628. https://doi.org/10.1016/j.jpowsour.2015.08.041

Liu S, Afacan A, Masliyah J (1994) Steady incompressible laminar flow in porous media. Chem Eng Sci 49:3565–3586. https://doi.org/10.1016/0009-2509(94)00168-5

Lohaus J, Rall D, Kruse M et al (2019) On charge percolation in slurry electrodes used in vanadium redox flow batteries. Electrochem Commun 101:104–108. https://doi.org/10.1016/j.elecom.2019.02.013

Mahdi RA, Mohammed HA, Munisamy KM, Saeid NH (2015) Review of convection heat transfer and fluid flow in porous media with nanofluid. Renew Sustain Energy Rev 41:715–734. https://doi.org/10.1016/j.rser.2014.08.040

John Newman NPB (2021) Electrochemical systems, 4th edn

Nield D, Bejan A (2006) Convection in porous media 3rd edn

Noack J, Roznyatovskaya N, Herr T, Fischer P (2015) The chemistry of redox-flow batteries. Angew Chemie Int Ed. https://doi.org/10.1002/anie.201410823

Park JY, Sohn DY, Choi YH (2020b) A numerical study on the flow characteristics and flow uniformity of vanadium redox flow battery flow frame. Appl Sci 10:1–17. https://doi.org/10.3390/app10238427

Park JY, Kim BR, Sohn DY et al (2020a) A study on flow characteristics and flow uniformity for the efficient design of a flow frame in a redox flow battery. Appl Sci. 10 https://doi.org/10.3390/app10030929

Prumbohm E, Wehinger GD (2019) Exploring flow characteristics in vanadium redox-flow batteries: optical measurements and CFD simulations. Chem-Ing-Tech 91:900–906. https://doi.org/10.1002/cite.201800164

Rychcik M, Skyllas-Kazacos M (1988) Characteristics of a new all-vanadium redox flow battery. J Power Sour 22:59–67. https://doi.org/10.1016/0378-7753(88)80005-3

Soloveichik GL (2015) Flow batteries: current status and trends. Chem Rev

Vafai K, Kim SJ (1995) On the limitations of the Brinkman-Forchheimer-extended Darcy equation. Int J Heat Fluid Flow 16:11–15. https://doi.org/10.1016/0142-727X(94)00002-T

Weller HG, Tabor G, Jasak H, Fureby C (1998) A tensorial approach to computational continuum mechanics using object-oriented techniques. Comput Phys 12:620. https://doi.org/10.1063/1.168744

Whitaker S (1996) The Forchheimer equation: a theoretical development. Transp Porous Media 25:27–61. https://doi.org/10.1007/BF00141261

Yang Z, Zhang J, Kintner-Meyer MCW et al (2011) Electrochemical energy storage for green grid. Chem Rev 111:3577–3613. https://doi.org/10.1021/cr100290v

Ye R, Henkensmeier D, Yoon SJ et al (2018) Redox flow batteries for energy storage: a technology review. J Electrochem Energy Conv Storage 15

Yoon SJ, Kim S, Kim DK (2019) Optimization of local porosity in the electrode as an advanced channel for all-vanadium redox flow battery. Energy 172:26–35. https://doi.org/10.1016/j.energy.2019.01.101

Zhang H, Li X, Zhang J (2017) Redox flow batteries: fundamentals and applications. Redox flow batteries: fundamentals and applications, pp 1–432

# Liquid Metal Batteries

**Norbert Weber and Tom Weier**

**Abstract** Liquid metal batteries (LMBs) are introduced as future candidates for grid scale electricity storage. Their completely liquid cell interior entails a prominent role of fluid mechanics to understand and model their behaviour. We describe the equations used to compute electrochemical reactions, heat and mass transfer, electromagnetic fields, and fluid flow and explain the simplifications that can be made in the case of LMBs. The implementation of solution algorithms in OpenFOAM pertaining to domain coupling, multiphase simulations, mesh mapping, and operator discretisation are discussed in detail and accompanied by example code.

## 1 Introduction

Liquid metal batteries (LMBs) are high temperature electricity storage devices. They consist of a low density molten alkaline or alkaline earth metal as the negative electrode (anode), a high density post-transition metal or metalloid as the positive electrode (cathode), and a fused salt of intermediate density as the ionic conductor. The three liquid layers arrange themselves into a stable density stratification as sketched in Fig. 1a by virtue of the immiscibility of the two metals with the salt.

LMBs possess a number of properties that make them attractive candidates for grid electricity storage. With the growing importance of the latter, research on LMBs that had ceased after an active period in the 1960s (Cairns and Shimotake 1969) was revived at the beginning of the 21st century (Kim et al. 2013). Current densities in LMBs are very high with typical values of around 1 A/cm$^2$. Structure degradation,

**Supplementary Information** The online version contains supplementary material available at https://doi.org/10.1007/978-3-030-92178-1_7.

N. Weber (✉) · T. Weier
Helmholtz-Zentrum Dresden - Rossendorf, 01328 Dresden, Germany
e-mail: norbert.weber@hzdr.de

T. Weier
e-mail: t.weier@hzdr.de

© Springer Nature Switzerland AG 2022

S. Beale and W. Lehnert (eds.), *Electrochemical Cell Calculations with OpenFOAM*, Lecture Notes in Energy 42, https://doi.org/10.1007/978-3-030-92178-1_7

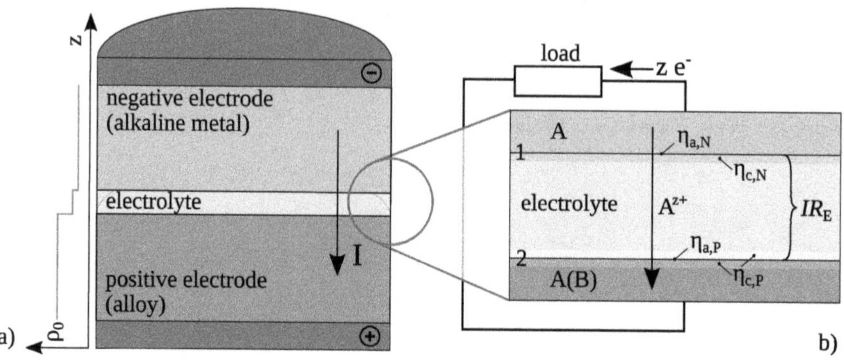


**Fig. 1** Sketch of a liquid metal battery with idealised density distribution at equilibrium (**a**) and zoom of the interfacial regions with typical overvoltages (**b**)

one of the major problems causing capacity fade and failure of batteries with solid electrodes, is absent due to the liquid state of the active materials. LMBs can be composed from a variety of elements allowing the selection of abundant and economic ones. The simple construction of the cells enables scaling on the cell level up to potentially very large dimensions like those common in aluminium refinement and electrolysis.

Fluid mechanics has a major role to play in the design and analysis of LMBs due to the fully liquid cell interior. Electric currents, magnetic fields, and heat and mass transfer are tightly coupled with the cells' electrochemistry. Fluid flow can be induced by a number of mechanisms with a subset compiled in Table 1. They are briefly touched upon in the following and discussed in detail by Kelley and Weier (2018).

Electric current flow leads to a strong heating of the electrolyte since its electrical conductivity is with $\mathcal{O}(100)$ S/m much smaller than that of the electrode metals with typically $\mathcal{O}(10^6)$ S/m. The heat source in the middle of the cell drives intense natural convection in the electrolyte itself and—to a lesser extent—in the negative electrode above (Shen and Zikanov 2016; Personnettaz et al. 2018). Marangoni convection due to temperature and concentration differences can be expected to occur mainly in the electrolyte at the interfaces with the electrodes (Köllner et al. 2017; Weier et al. 2017).

At discharge, the anode metal is alloyed into the positive electrode; during charging, this process is reversed. The large density contrast between the anode and cathode metals causes strong density gradients in the positive electrode during charge and discharge. These gradients are stable during discharge but unstable during charge, inhibiting or driving solutal convection in the positive electrode, respectively (Personnettaz et al. 2019; Herreman et al. 2020; Personnettaz et al. 2020).

Finally, electromagnetic effects of the strong total currents that result from the high current densities in large cells can cause fluid flow. The Tayler instability (TI) (Weber et al. 2013) is a kink-type instability akin to the z-pinch known from plasma physics.

**Table 1** Convection types to be expected in liquid metal batteries

| Convection type | Typical location | Cause |
|---|---|---|
| Tayler instability | Negative electrode | Electromagnetic |
| Metal pad role instability | Interfaces | Electromagnetic |
| Electro-vortex flows | Electrodes | Electromagnetic |
| heat driven convection | Electrolyte | Temperature gradient |
| | Negative electrode | Temperature gradient |
| Marangoni convection (thermal) | Electrolyte at both interfaces | Temperature gradient |
| | Electrodes at both interfaces | Temperature gradient |
| Marangoni convection (solutal) | Electrolyte at both interfaces | Concentration gradient |
| | Positive electrode at interface | Concentration gradient |
| Solutal convection | Positive electrode | Concentration gradient |

For total currents above a few kA the TI will occur in the negative electrode first due to its material properties. While even perfectly homogeneous current density distributions are subject to the TI, changes in a conductors cross section that are technically unavoidable lead to radial current density components and rotational Lorentz forces. The latter produce so-called electro-vortex flows. These flows might be weak, but there is no need to exceed a threshold current for them to be present.

The two interfaces between electrodes and electrolyte may develop interfacial waves (Weber et al. 2017) similar to those observed in aluminium reduction cells (ARCs). Large amplitude waves could short circuit the cell and should therefore be avoided. Unlike ARCs, LMBs posses two fluid-fluid interfaces that may interact, leading to a richer dynamics (Horstmann et al. 2017) compared to the single interface case.

The cell voltage $E$ of an LMB under current flow $I$ depends on the open circuit voltage $E_{\mathrm{OC}}$ and the sum of overvoltages $\eta$:

$$E = E_{\mathrm{OC}} \pm (I R_{\mathrm{E}} + \eta_{\mathrm{a,N}} + \eta_{\mathrm{c,N}} + \eta_{\mathrm{a,P}} + \eta_{\mathrm{c,P}}), \tag{1}$$

with the activation overvoltages (subscript a) $\eta_{\mathrm{a,N}}$, $\eta_{\mathrm{a,P}}$ at the positive (subscript P) and negative (subscript N) electrodes. The corresponding mass transfer overvoltages (subscript c) are denoted by $\eta_{\mathrm{c,N}}$, $\eta_{\mathrm{c,P}}$ and the ohmic voltage loss is due to the resistance of the electrolyte $R_{\mathrm{E}}$. The plus sign in Eq. (1) applies to charge, the minus sign to discharge. The open circuit voltage itself

$$E_{\mathrm{OC}} = -\frac{RT}{zF} \ln a_{\mathrm{A(B)}} \tag{2}$$

depends on the activity $a_{\mathrm{A(B)}}$ of the negative electrode material A in the positive electrode material B. $R$ is the ideal gas constant, $T$ the temperature, $z$ the valence, and $F$

the Faraday constant (see Sect. 2.1 for details). Charge transfer at the liquid-liquid interface is very fast at the high operating temperatures (300–600 °C). Therefore the activation overvoltages are negligible. The electrolyte is the source of the ohmic voltage drop $IR_{\rm E}$ that is in most cases the largest contribution to the total overvoltage. Mass transfer limitations in the electrolyte may arise, if salt mixtures with different cations are used. However, this brings about a number of other detrimental effects and is therefore avoided in most cases. Irrespective of the salt mixture, heat release in the electrolyte should generate intense convection and sufficient mixing so that no substantial mass transfer overvoltages are to be expected in the electrolyte. The situation is, however, different for the positive electrode. Since it is the activity of A(B) directly at the interface to the electrolyte that determines $E_{\rm OC}$, mass transfer in the positive electrode is crucial for cell performance. Comparing the contribution of solutal and thermal gradients to the alloy density differences reveals (Kelley and Weier 2018; Personnettaz et al. 2019) that the influence of compositional gradients far outweighs that of temperature. As mentioned above, strong solutal convection appears during charge, but diffusion dominates mass transfer during discharge and limits cell performance. The aforementioned electro-vortex flows are a good candidate to counteract this stable density stratification and to improve mixing in the positive electrode (Weber et al. 2018) during discharge.

## 2 Physical and Numerical Model

### *2.1 Electrochemistry*

When discharging an LMB, metal A is oxidised at the anode-electrolyte interface as

$$\mathrm{A} \longrightarrow \mathrm{A}^{z+} + \mathrm{ze}^-,$$

crosses the electrolyte and is reduced at the cathode-electrolyte interface as

$$\mathrm{A}^{z+} + \mathrm{ze}^- \longrightarrow \mathrm{A(B)}$$

where it dissolves into metal B (Kim et al. 2013). For clarity, we will focus in the following on the well investigated Li||Bi cell, where Li dissolves into Bi. There, the number of exchanged electrons is $z = 1$.

From a macroscopic point of view, the electric potential will jump at both interfaces of an LMB—as illustrated in Fig. 2.

The cell current will always drive flow in the liquid electrodes—e.g. by heating or electromagnetic forces (Ashour et al. 2018; Weber et al. 2018). Therefore, it is not sufficient to model only the scalar value of the cell current: we need to know its three-dimensional distribution, as well (Weber et al. 2019, 2020). This is quite

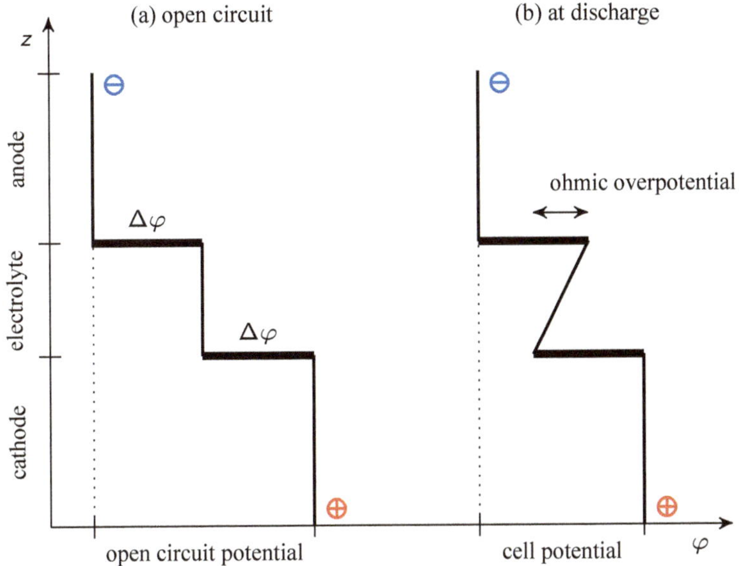


**Fig. 2** Vertical profile of the electric potential in an LMB at open circuit (**a**) and discharge (**b**) (Weber et al. 2019)

remarkable, because for classic batteries it is sufficient to find the overpotentials in, or at the electrolyte—but not the current distribution in the electrodes.

In a first step, the electric potential $\varphi$ will be solved as

$$\nabla \cdot \sigma \nabla \varphi = 0, \tag{3}$$

where $\sigma$ denotes the ohmic or ionic conductivity. The potential jumps at the two interfaces are embedded into the Laplace operator, and are defined by the Nernst equation as

$$\Delta \varphi = -\frac{RT}{zF} \ln \left( \frac{a_{\text{ox}}}{a_{\text{red}}} \right), \tag{4}$$

with $R$, $T$, $F$ and $a$ denoting the Universal gas constant, temperature, Faraday constant and chemical activity of the oxidated and reduces species. Finally, the current density $\boldsymbol{j}$ is computed as

$$\boldsymbol{j} = -\sigma \nabla \varphi. \tag{5}$$

Note that LMBs are simple concentration cells. Their cell voltage is therefore simply defined by the activity of one metal in the other metal.

**Practical Advice**

In a Li|LiCl-LiF|Bi LMB, the activity of $Li^+$ in LiCl-LiF is not known. In that case, the potential jump might be applied only at the bottom interface, computing its magnitude as $\Delta\varphi = -\frac{RT}{zF} \ln\left(a_{\mathrm{Li(Bi)}}\right)$. Even simpler, measured values of the open-circuit potential of a concentration cell can be fitted over concentration, and its value applied directly as potential jump at the electrolyte-cathode interface.

## 2.2 *Mass Transfer*

The potential jump at the interface depends on the activity—or concentration—of Li in Bi. Consequently, we need to model how Li dissolves into Bi in order to compute the cell voltage. We solve the diffusion-advection equation (Personnettaz et al. 2019, 2020)

$$\frac{\partial}{\partial t}\gamma + \nabla \cdot (\boldsymbol{u}\gamma) = \nabla \cdot (D\nabla\gamma) \tag{6}$$

for the mass concentration $\gamma$ of Li, with $t$ denoting time, $\boldsymbol{u}$ velocity and $D$ the diffusion coefficient of Li in Bi. Although the density difference between Li and Bi is extreme, the Oberbeck-Boussinesq approximation is typically used to model solutal convection in LMBs. The body force caused by the concentration (i.e. density) gradients may be computed as

$$\boldsymbol{f} = \rho\boldsymbol{g} = \rho_0(1 - \beta_\gamma(\gamma - \gamma_0))\boldsymbol{g}, \tag{7}$$

where $\gamma_0$ denotes a reference concentration, $\boldsymbol{g}$ gravity, $\rho_0$ the reference density and $\beta_\gamma$ the volumetric expansion coefficient. The amount of Li entering the positive electrode is proportional to the current density, and can be computed using Faraday's law. The normal gradient of the Li mass concentration at the interface is then

$$\nabla\gamma \cdot \boldsymbol{n} = -\frac{\boldsymbol{j}M}{zFD} \cdot \boldsymbol{n} \tag{8}$$

with $M$ denoting the molar mass of Li and $\boldsymbol{n}$ the surface normal vector. Typical current densities for LMBs are in the order of 0.2-0.3 A/cm$^2$, but may reach even 1 A/cm$^2$.

**Mass vs. Molar Concentration and Fraction**

It is very much recommended to use mass concentration as primary variable for the diffusion-advection equation. In its original form, Fick's second law of diffusion reads

$$\frac{\partial c}{\partial t} = D\Delta c, \tag{9}$$

with $c$ denoting the molar concentration in mol/m$^3$. Multiplying Fick's law by the molar mass of Li leads to

$$\frac{\partial \gamma}{\partial t} = D\Delta \gamma. \tag{10}$$

The mass concentration $\gamma$ can be expressed by the mass fraction $w$ and the density of the mixture $\rho$ as $\gamma = w \cdot \rho$ which leads using the product rule to

$$\rho \frac{\partial w}{\partial t} + w \frac{\partial \rho}{\partial t} = D\rho\Delta w + Dw\Delta\rho. \tag{11}$$

As the density changes considerably when alloying Li into Bi, the terms $w\frac{\partial \rho}{\partial t}$ and $Dw\Delta\rho$ are not zero!

The Navier-Stokes equations describe how fast mass is transported, i.e. we use mass-averaged velocity. Alternatively, we could define a molar-averaged velocity, describing how fast the *amount* of Li and Bi is transported. When adding Li into Bi, the total amount, but also the mass of the mixture will change. If the change of the amount of substance and the density are linearly related to each other, the mass- and molar-averaged velocity are equal. This is, however, not the case for liquid metal batteries: when adding Li to Bi, the small Li atoms will intercalate between the large Bi atoms. While the total amount of substance changes considerably, density changes only slightly. Consequently, we need to use mass concentration as primary variable, because our velocity is always mass-averaged (Bird et al. 1960).

## 2.3 Flow Simulation

Convection in the liquid phases is modelled by solving the incompressible Navier-Stokes equations (Weber et al. 2018)

$$\frac{\partial(\rho \boldsymbol{u})}{\partial t} + \nabla \cdot (\rho \boldsymbol{u}\boldsymbol{u}) = -\nabla p + \nabla \cdot (\rho\nu(\nabla \boldsymbol{u} + (\nabla \boldsymbol{u})^{\mathsf{T}})) + \sum \boldsymbol{f}, \tag{12}$$

$$\nabla \cdot \boldsymbol{u} = 0, \tag{13}$$

with $p$ denoting the pressure, $\nu$ the kinematic viscosity and $\boldsymbol{f}$ an arbitrary body force, such as gravitation or a Lorentz force. The equation system is solved using the PISO algorithm by first estimating a velocity. After solving a Poisson equation for the pressure, the velocity is corrected ensuring $\nabla \cdot \boldsymbol{u} = 0$. In certain cases, such as multiphase simulation or with thermal convection, it is better to switch off the momentum predictor in the *fvSolution* dictionary. Then, the velocity of the old time step is used directly as estimate for the new velocity. The corrector step, i.e. the Poisson equation for pressure, should always be solved at least twice in order to ensure convergence.

**Simplifications for Single Phase Flow**

When simulating a single phase, the equation is usually divided by the constant density $\rho$. Assuming further the kinematic viscosity to be constant, the stress tensor can be simplified as

$$\nabla \cdot \rho\nu \left(\nabla \boldsymbol{u} + \nabla \boldsymbol{u}^{\mathsf{T}}\right) = \nabla \cdot (\rho\nu\nabla \boldsymbol{u}) + \nabla \boldsymbol{u} \cdot \nabla (\rho\nu) = \nabla \cdot (\rho\nu\nabla \boldsymbol{u}),$$

because $\nabla(\rho\nu) = 0$.

## 2.4 *Heat Transfer and Thermal Convection*

The temperature distribution in the cell is determined by solving the energy equation

$$c_p \left( \frac{\partial \rho T}{\partial t} + \nabla \cdot (\rho T \boldsymbol{u}) \right) = \nabla \cdot \lambda \nabla T + \frac{j^2}{\sigma} + \dot{Q}, \tag{14}$$

with $c_p$ denoting the isobaric heat capacity, $\lambda$ the thermal conductivity and $j^2/\sigma$ the ohmic heat source. Additional heat sources $\dot{Q}$—such as the electrochemical heat due to the reaction—are sometimes included, as well (Personnettaz et al. 2018).

Thermal convection is modelled using the Oberbeck-Boussinesq approximation. The gravitational force is defined as

$$\boldsymbol{f} = \rho \boldsymbol{g} = \rho_0 (1 - \beta (T - T_0)) \boldsymbol{g}, \tag{15}$$

with $\beta$ denoting the thermal expansion coefficient and $T_0$ the reference temperature. The part of the force which can be expressed as a gradient does not drive a flow. In order to reduce numerical errors, it is therefore included into the pressure gradient by defining a modified pressure as (Rusche 2002)

$$p_{\mathrm{d}} = p - \rho \boldsymbol{g} \cdot \boldsymbol{x}, \tag{16}$$

where $\boldsymbol{x}$ denotes the position vector. The gradient of the pressure becomes then

$$\nabla p = \nabla p_{\mathrm{d}} + \rho \boldsymbol{g} + \boldsymbol{g} \cdot \boldsymbol{x} \nabla \rho, \tag{17}$$

and $\rho \boldsymbol{g}$ disappears from the Navier-Stokes equation.

**The Oberbeck-Boussinesq Aroximation**

The Oberbeck-Boussinesq approximation is based on three assumptions (Gray and Giorgini, 1976):

- heating due to viscous dissipation is assumed to be negligible
- all material properties except density do not depend on temperature
- density changes are solely taken into account in the gravitational term

After choosing an allowed error $\varepsilon$, the admissible temperature difference in the simulation can roughly be estimated as

$$\Delta T = \frac{\varepsilon}{\beta}. \tag{18}$$

---

## 2.5 *Magnetohydrodynamics*

Magnetohydynamic effects can drive fluid flow by the Lorentz force

$$\boldsymbol{f} = \boldsymbol{j} \times \boldsymbol{b}, \tag{19}$$

i.e. by the cross-product of the current density and a magnetic field. The current density is determined – similar to Eqn. 3 – by first solving the Poisson equation

$$\nabla \cdot \sigma \nabla \varphi = \nabla \cdot \sigma (\boldsymbol{u} \times \boldsymbol{B}), \tag{20}$$

and then computing

$$\boldsymbol{j} = -\sigma \nabla \varphi + \sigma (\boldsymbol{u} \times \boldsymbol{B}) - \sigma \frac{\partial \boldsymbol{a}}{\partial t}, \tag{21}$$

with $\boldsymbol{a}$ denoting the vector potential. Simulating only direct currents, the last term of Eq. 21 can safely be neglected (Weber et al. 2013). The magnetic field may then be obtained by solving the quasi-static induction equation

$$0 = \frac{1}{\sigma \mu_0} \Delta \boldsymbol{b} + \nabla \times (\boldsymbol{u} \times \boldsymbol{b}) \tag{22}$$

with $\mu_0$ denoting the vacuum permeability. The boundary conditions for the magnetic field are obtained from Biot-Savart's law as (Weber et al. 2015)

$$\boldsymbol{b}(\boldsymbol{r}) = \frac{\mu_0}{4\pi} \int \frac{\boldsymbol{j}(\boldsymbol{r}') \times (\boldsymbol{r} - \boldsymbol{r}')}{|\boldsymbol{r} - \boldsymbol{r}'|^3} dV', \tag{23}$$

where $\boldsymbol{r}$ denotes the coordinate vector of the magnetic field, $\boldsymbol{r}'$ of the current density and $dV'$ the cell volume.

Alternatively, a transport equation for the vector potential may be solved as

$$0 = \Delta \boldsymbol{a} + \mu \boldsymbol{j} \tag{24}$$

with the boundary conditions obtained by Green's identity (Weber et al. 2018)

$$\boldsymbol{a}(\boldsymbol{r}) = \frac{\mu_0}{4\pi} \int \frac{\boldsymbol{j}(\boldsymbol{r}')}{|\boldsymbol{r} - \boldsymbol{r}'|} dV'. \tag{25}$$

Finally, the magnetic field is obtained as

$$\boldsymbol{b} = \nabla \times \boldsymbol{a}. \tag{26}$$

**Practical Advice**

Usually, large parts of the Lorentz force can be expressed as a pressure gradient, and therefore drive no flow. However, these large gradients can lead to considerable numerical errors. It is therefore highly recommended to avoid non-orthogonal cells. Computing the magnetic field directly—and not via the vector potential—is slightly slower. However, it is more accurate, because we do not need to compute an additional rotation ($\boldsymbol{b} = \nabla \times \boldsymbol{a}$). Finally, those parts of the Lorentz force, which can be expressed as a pressure gradient, should whenever possible be included into a modified pressure.

## 3 Domain Coupling

In many cases simulating a single electrode of the LMB is sufficient. Then, the equations can be considerably simplified, e.g. by diving them by the constant material properties. However, in certain cases it is necessary to simulate the full battery. Coupling non-deformable regions, like the battery housing with a liquid electrode, we will denote as "region coupling" in the following. Multiphase models describe how to simulate the interaction of the three deformable liquid layers as illustrated in Fig. 3.

### *3.1 Multiphase Simulation*

Multiphase simulations are required to model the deformation of the interfaces between electrodes and electrolyte. The solver is based on the OpenFOAM standard solver *multiphaseInterFoam* using the volume of fluid method (Rusche 2002).

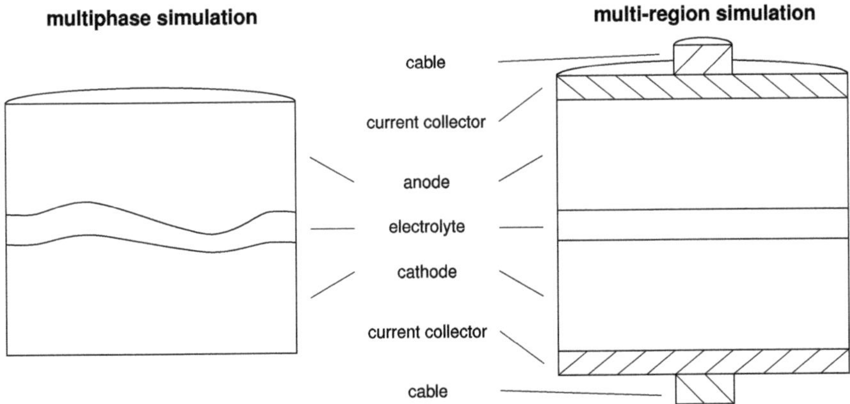


**Fig. 3** While multiphase simulation can model the deformation of the interfaces between the electrodes and the electrolyte (left), multi-region modelling allows coupling of the current collectors and cables with the battery (right)

The volumetric phase fractions $\alpha_i$ describe the volume of a single phase in each computational cell. After solving the Navier-Stokes equations, the phase fractions $\alpha_i$ are solved as

$$\frac{\partial \alpha_i}{\partial t} + \nabla \cdot (\alpha_i \boldsymbol{u}) = 0. \tag{27}$$

Thereafter, the mixture properties can be found as (Weber et al. 2017; Horstmann et al. 2017; Personnettaz et al. 2018)

$$\nu = \frac{1}{\rho}\sum_i \alpha_i \rho_i \nu_i, \quad c_p = \frac{1}{\rho}\sum_i \alpha_i \rho_i c_{p,i},$$
$$\lambda = \left(\sum_i \frac{\alpha_i}{\lambda_i}\right)^{-1}, \quad \sigma = \left(\sum_i \frac{\alpha_i}{\sigma_i}\right)^{-1} \quad \text{and} \quad \rho = \sum_i \alpha_i \rho_i. \tag{28}$$

Note that the kinematic viscosity $\nu$ and heat capacity $c_p$ are weighted by density, the thermal conductivity $\lambda$ and electric conductivity $\sigma$ harmonically and the density linearly, as illustrated schematically in Fig. 4. Harmonic weighting is especially important for electric conductivity, because its value changes typically by four orders of magnitude between the electrodes and the molten salt.

Surface tension is added as a volumetric force around the interface as

$$f = \sum_i \sum_{j \neq i} \gamma_{ij} \kappa_{ij} \delta_{ij} \tag{29}$$

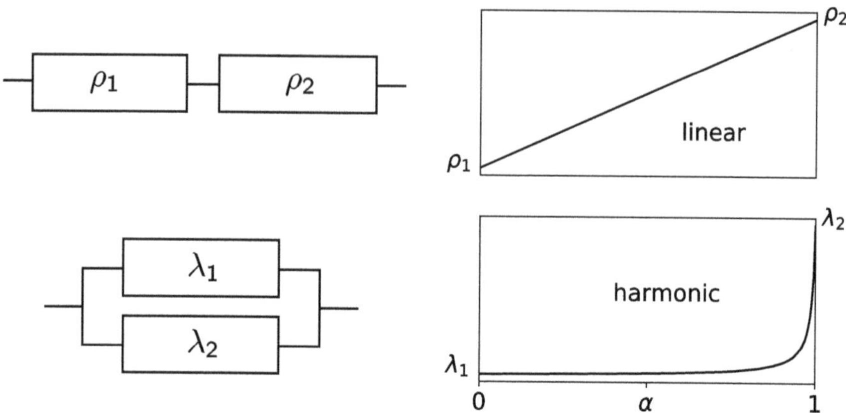


**Fig. 4** Densities are weighted linearly like two resistances in a serial circuit, while thermal and electrical conductivities are harmonically weighted as resistivities in a parallel circuit

using the CSF model of Brackbill et al. (1992). Here, $\gamma_{ij}$ denotes the surface tension between phase $i$ and $j$ with the curvature of the interface defined as

$$\kappa_{ij} = -\nabla \cdot \frac{\alpha_j \nabla \alpha_i - \alpha_i \nabla \alpha_j}{|\alpha_j \nabla \alpha_i - \alpha_i \nabla \alpha_j|}. \tag{30}$$

Finally, the term $\delta_{ij} = \alpha_j \nabla \alpha_i - \alpha_i \nabla \alpha_j$ ensures that the surface tension force is applied only near the interface.

## Spurious Velocities

Spurious velocities are unphysical velocities that appear in volume of fluid simulations near the interface—and can easily reach 1 cm/s in LMB simulations, if no countermeasures are taken. Generally, spurious velocities are caused by an imbalance of the pressure gradient with an arbitrary volume force. Typical sources for spurious currents are large density jumps between two phases, a bad curvature calculation and the explicit discretisation of the surface tension force. Very simple measures for reducing spurious currents include lowering the pressure residual, and basing the time step on the capillary Courant number as (Personnettaz et al. 2018)

$$\Delta t = \sqrt{\frac{(\rho_A + \rho_B)\Delta x^3}{2\pi \gamma_{\max}}} \cdot \text{Co}_{\text{cap}}, \tag{31}$$

with $\rho_A$ and $\rho_B$ denoting the densities of two phases, $\Delta x$ the mesh cell size and $\gamma_{\max}$ the largest interface tension.

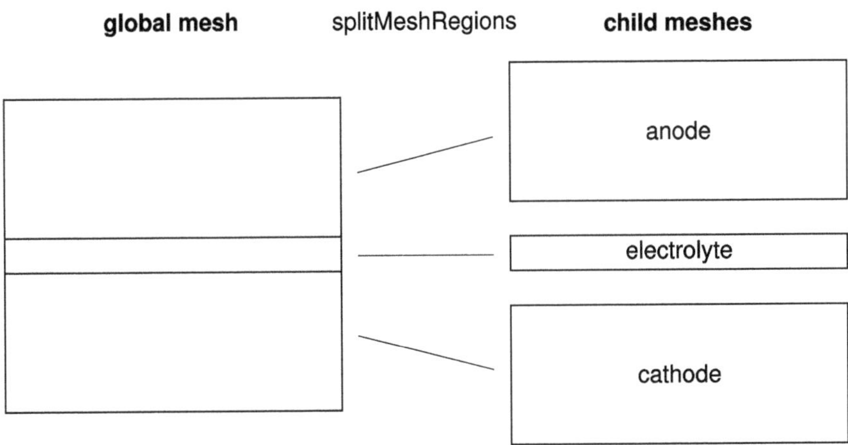


**Fig. 5** The OpenFOAM utility splitMeshRegion is used to split the global mesh into several child meshes

## 3.2 *Multi-Region Simulation*

When using the multi-region approach, the interfaces between electrodes and electrolyte are assumed to be rigid. Different equations are solved in different regions, e.g. one in the electrode and another in the electrolyte. The meshes are coupled at the boundary, or in the volume by mesh-to-mesh interpolation.

### 3.2.1 Parent-Child Mesh Method

Certain variables, such as temperature, electric potential or current density, exist in the whole battery. It is therefore opportune to solve them on a global mesh covering all the LMB. Material properties are then simply defined as a *volScalarField* to account for their changes between different conductors.

Distinct from the global mesh, one child mesh is defined for each single region. This allows solving local variables (as e.g. the concentration) in the appropriate region only. Local source terms, such as the heat of reaction, may be mapped easily between child and parent mesh, as both are overlapping.

On a first glance the handling of, and the interpolation between different meshes seems to be complicated. However, the parent-child mesh technique has one important advantage: simulation is fast. Renouncing to the global mesh would require solving e.g. temperature on each region-mesh. This can be very slow, because the fields are only coupled at the boundary. Especially when using explicit coupling, one would need to iterate between the different regions for many times up to convergence.

Still, one notable exception exists: velocity. Solving for velocity and pressure on one global mesh—with rigid boundaries—would require defining an appropriate coupling for pressure and velocity. As that might be complex, it is better to solve the flow on the child meshes, and couple the velocity at the interfaces.

### 3.2.2 Mapping Between Meshes

The global mesh is split into child meshes by running the OpenFOAM utility *splitMeshRegions* as illustrated in Fig. 5. This creates three files for each child: faceRegionAddressing, cellRegionAddressing and boundaryRegionAddressing. Here, boundaryRegionAddressing is a simple list, where entry $i$ contains the parent-boundary number which belongs to the child-patch $i$. Similarly, the entry $i$ in cellRegionAddressing gives the parent-cell number, belonging to child-cell $i$. The file faceRegionAddressing works similarly, but contains additional information: the orientation of the faces. If an entry of faceRegionAddressing is positive, the parent and child face have the same orientation—otherwise they are inverse. As zero can not have a sign, faceRegionAddressing counts beginning from 1. This means, we obtain the parent-face number of child-face $i$ by

```
label parentFace = mag(fluidFaceRegionAddressing[i]) - 1;
```

and its relative orientation by

```
scalar orientation = sign(fluidFaceRegionAddressing[i]);
```

Within a multi-region solver, we read the three files mentioned above for each child-mesh, and save the mapping information as a cellMap, boundaryMap and faceMap, and the orientation of the faces as a faceMask—for an illustration, see Fig. 6.

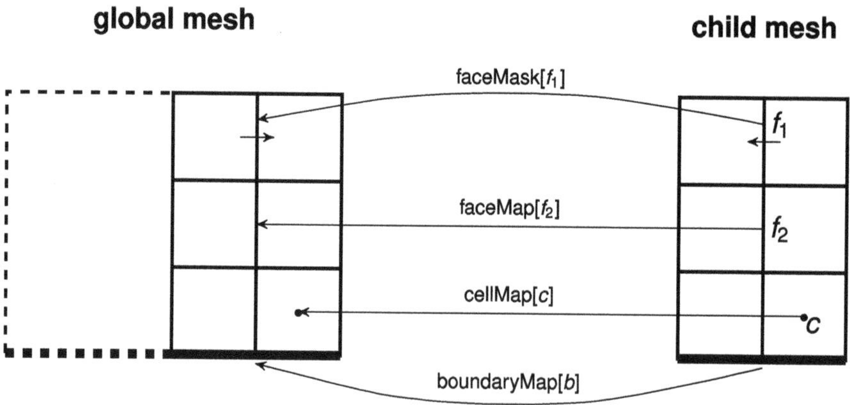


**Fig. 6** Mapping between the global and a single child mesh

**Child to parent mapping**
With this information, cell values can simply be mapped from a child mesh to the parent mesh by invoking

```
globalField.rmap(childField, cellMap);
```

Similarly, patchI can be mapped from the child to the parent-mesh by first defining a local field mapper as

```
labelField tmpMap
(
    labelField::subField
    (
        faceMap,
        childMesh.boundary()[patchI].size(),
        childMesh.boundary()[patchI].patch().start()
    )
);
```

and subtracting then the start index of the parent-patch as

```
tmpMap -= globalMesh.boundary()[patchesMap[patchI]].patch().start();
```

Finally, the boundary field is mapped as

```
globalField.boundaryFieldRef()[patchesMap[patchI]].
      scalarField::rmap(childField.boundaryField()[patchI], tmpMap);
```

**Parent to child mapping**

Mapping cell values from the parent to a child mesh is similarly easy:

```
forAll(childField, cellI)
{
    childField[cellI] = globalField[cellMap[cellI]];
}
```

Boundary values are mapped analogue as described before by defining a local field mapper, subtracting the start index of the parent mesh and invoking then

```
forAll(childField.boundaryField()[patchI], faceI)
{
    childField.boundaryFieldRef()[patchI][faceI] =
        parentField.boundaryField()[patchesMap[patchI]][tmpMap[faceI]];
}
```

## 4 Discretisation

The standard discretisation schemes are used for most equations. Only the exceptions are described below: the gradient and Laplace operator for the electric potential, the interpolation of temperature and potential as well as the interpolation of thermal and electrical conductivity need special attention.

## 4.1 Laplace Operator

When solving the Laplace equation for the electric potential, internal voltage jumps $\Delta\varphi$ need to be accounted for as described in Sect. 2.1. The Laplace operator is discretised using the Gauss theorem as

$$\nabla \cdot \sigma \nabla \varphi = \sum_f \boldsymbol{S} \sigma_{\mathrm{f}} (\nabla \varphi)_{\mathrm{f}} = \sum_f |\boldsymbol{S}| \sigma_{\mathrm{f}} \frac{\varphi_{\mathrm{N}} - \varphi_{\mathrm{P}} + \Delta\varphi}{|\boldsymbol{d}|}, \tag{32}$$

where $\boldsymbol{S}$ denotes the face-normal vector, $\sigma_{\mathrm{f}}$ the conductivity on the face, $(\nabla\varphi)_{\mathrm{f}}$ the potential gradient on the face, $\varphi_{\mathrm{P}}$ the potential in the cell centre of the owner cell, $\varphi_{\mathrm{N}}$ the potential in the cell centre of the neighbour cell and $|\boldsymbol{d}|$ the distance between both cells. Compared to the standard discretisation, only the potential jump $\Delta\varphi$ is added. The jump is defined as a *surfaceScalarField*, which is zero on all faces except at the interface.

## 4.2 Gradient Operator

The gradient operator for the electric potential needs special attention, because it must account for the jumps at the interfaces. It is discretised as (Jasak 1996)

$$\nabla \varphi = \frac{1}{V} \sum_f \boldsymbol{S} \varphi_{\mathrm{f}}, \tag{33}$$

with $V$ denoting the cell volume, and the electric potential at the faces defined as

$$\varphi_{\mathrm{fP}} = w \cdot (\varphi_{\mathrm{N}} - \Delta\varphi) + (1 - w) \cdot \varphi_{\mathrm{P}}, \tag{34}$$

and the weighting factor

$$w = \frac{\delta_{\mathrm{P}} \cdot \sigma_{\mathrm{N}}}{\delta_{\mathrm{N}} \sigma_{\mathrm{P}} + \delta_{\mathrm{P}} \sigma_{\mathrm{N}}}. \tag{35}$$

Here, $\delta_{\mathrm{P}}$ and $\delta_{\mathrm{N}}$ denote the distance between face and cell centre for the owner and neighbour cell, as illustrated in Fig. 7.

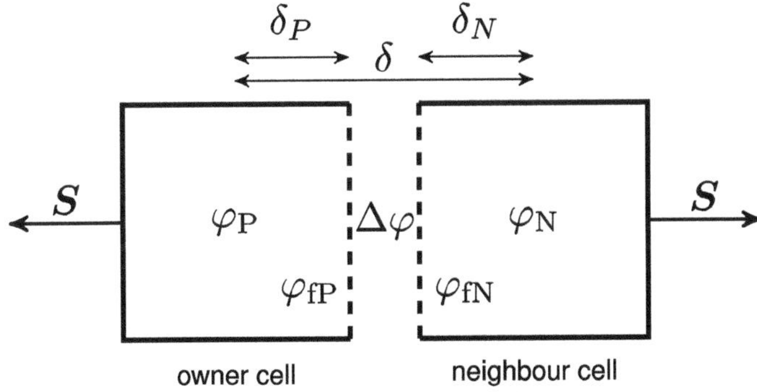


**Fig. 7** Discretisation of the electric potential for the Laplace and gradient operator

## 4.3 *Interpolation Schemes*

### 4.3.1 Thermal and Electric Conductivity

When discretising the Laplace equation (Eq. 32), i.e. $\nabla \cdot (\sigma \nabla \varphi) = 0$, the electrical conductivity needs to be interpolated from cell centres to the faces. While linear interpolation can be used for resistance, conductivity needs to be harmonically interpolated as (Weber et al. 2018)

$$\sigma_{\mathrm{f}} = \frac{1}{\dfrac{w}{\sigma_{\mathrm{P}}} + \dfrac{1-w}{\sigma_{\mathrm{N}}}}, \tag{36}$$

with P denoting the owner cell and N the neighbour cell. The weighting factor reads

$$w = \frac{\delta_{\mathrm{P}}}{\delta_{\mathrm{P}} + \delta_{\mathrm{N}}} \tag{37}$$

with $\delta_{\mathrm{P}}$ denoting the distance from the cell centre to the face of the owner, and similar for $\delta_{\mathrm{N}}$ of the neighbour cell (see Fig. 8).

### 4.3.2 Temperature and Electric Potential

When computing the current density as $\boldsymbol{j} = -\sigma \nabla \varphi$, the electric potential needs to be interpolated from cell centres to the faces. As the local potential depends strongly

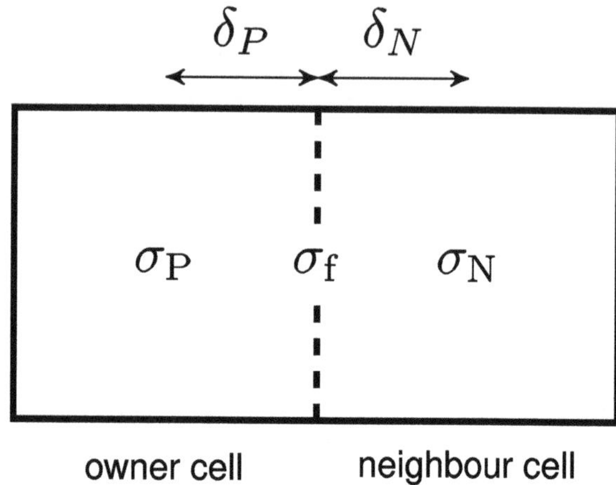


**Fig. 8** Interpolation of the electric and thermal conductivity from the cell centres to the faces

on the electric conductivity, the potential needs to be weighted by conductivity when being interpolated. We find the potential at the face as (Weber et al. 2018)

$$\varphi_\mathrm{f} = w\varphi_\mathrm{P} + (1 - w)\varphi_\mathrm{N} \tag{38}$$

with

$$w = \frac{\delta_\mathrm{N}\sigma_\mathrm{P}}{\delta_\mathrm{P}\sigma_\mathrm{N} + \delta_\mathrm{N}\sigma_\mathrm{P}}. \tag{39}$$

## 5 Example

In this section we illustrate the application of our models with a simple example. The corresponding source code is provided with the book. We model the solutal convection in the cathode of an LMB during charge. We solve the Navier-Stokes equation (Eq. 12) for a single fluid with one single body force: the buoyancy due to concentration gradients (Eq. 7). We find the concentration of Li in Bi by solving the diffusion-convection equation (Eq. 6) using Faraday's law as boundary condition (Eq. 8).

During the first three seconds, Li only diffuses out of the electrode (Fig. 9). However, after only 7.5 seconds small fingers of heavy Bi start to sink down reaching the bottom of the cell in only 15 seconds.

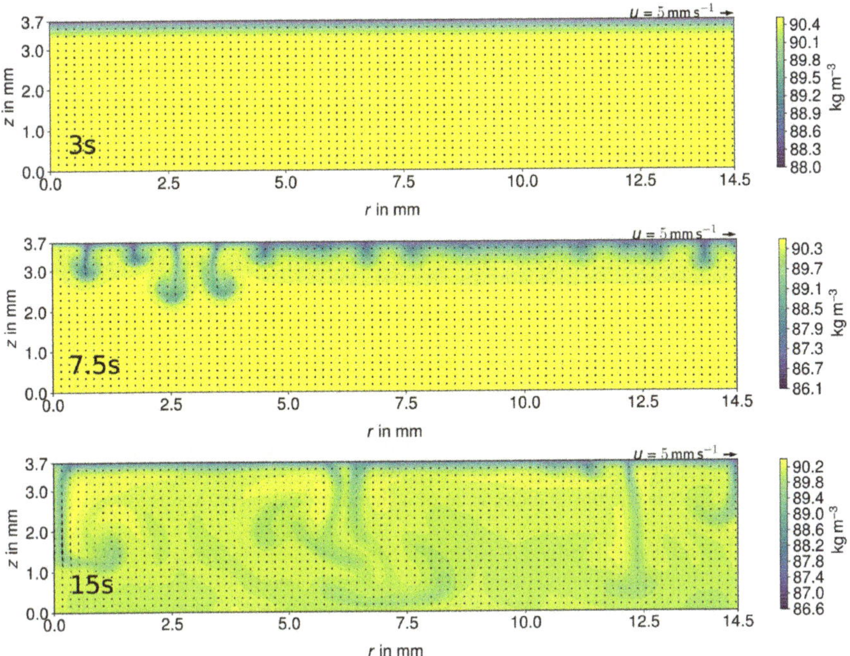


**Fig. 9** Mass concentration and velocity vectors in a LiBi cathode during charge of an LMB

**Acknowledgements** This project has received funding from the European Union's Horizon 2020 research and innovation programme under grant agreement No 963599, and was supported by the Deutsche Forschungsgemeinschaft (DFG, German Research Foundation) by award number 338560565, by a postdoc fellowship of the German Academic Exchange Service (DAAD) and in frame of the Helmholtz—RSF Joint Research Group "Magnetohydrodynamic instabilities: Crucial relevance for large scale liquid metal batteries and the sun-climate connection", contract No. HRSF-0044 and RSF-18-41-06201. We thank P. Personnettaz for providing the source code and example.

## References

Ashour RF, Kelley DH, Salas A, Starace M, Weber N, Weier T (2018) Competing forces in liquid metal electrodes and batteries. J. Power Sources 378:301–310

Bird RB, Stewart WE, Lightfoot EN (1960) Transport Phenomena. John Wiley and Sons, New York

Brackbill JU, Kothe DB, Zemach C (1992) A continuum method for modeling surface tension. J. Comput. Phys. 100:335–354

Cairns EJ, Shimotake H (1969) High-temperature batteries. Science 164:1347–1355

Horstmann GM, Weber N, Weier T (2018) Coupling and stability of interfacial waves in liquid metal batteries. *J. Fluid Mech.*, 845:1–35

Gray DD, Giorgini A (1976) The validity of the Boussinesq approximation for liquids and gases. Int. J. Heat Mass Transf. 19(5):545–551

Rusche H (2002) *Computational Fluid Dynamics of Dispersed Two-Phase Flows at High Phase Fractions*. PhD thesis, Imperial College London

Herreman W, Bénard S, Nore C, Personnettaz P, Cappanera L, Guermond J-L (2020) Solutal buoyancy and electrovortex flow in liquid metal batteries. Phys. Rev. Fluids 5(7):074501

Jasak H (1996) *Error Analysis and Estimation for the Finite Volume Method with Applications to Fluid Flows*. PhD thesis, Imperial College of Science, Technology and Medicine

Kelley DH, Weier T (2018) Fluid mechanics of liquid metal batteries. Appl. Mech. Rev. 70(2):020801

Kim H, Boysen DA, Newhouse JM, Spatocco BL, Chung B, Burke PJ, Bradwell DJ, Jiang K, Tomaszowska AA, Wang K, Wei W, Ortiz LA, Barriga SA, Poizeau SM, Sadoway DR (2013) Liquid metal batteries: past, present, and future. Chem. Rev. 113(3):2075–2099

Köllner T, Boeck T, Schumacher J (2017) Thermal Rayleigh-Marangoni convection in a three-layer liquid-metal-battery model. Phys. Rev. E 95:053114

Weber N, Beckstein P, Herreman W, Horstmann GM, Nore C, Stefani F, Weier T (2017) Sloshing instability and electrolyte layer rupture in liquid metal batteries. *Phys. Fluids*, 29(5):054101

Personnettaz P, Landgraf S, Nimtz M, Weber N, Weier T (2019) Mass transport induced asymmetry in charge/discharge behavior of liquid metal batteries. Electrochem. Commun. 105:106496

Personnettaz P, Landgraf S, Nimtz M, Weber N, Weier T (2020) Effects of current distribution on mass transport in the positive electrode of a liquid metal battery. Magnetohydrodynamics 56(2/3):247–254

Personnettaz P, Beckstein P, Landgraf S, Köllner T, Nimtz M, Weber N, Weier T (2018) Thermally driven convection in Li||Bi liquid metal batteries. J. Power Sources 401:362–374

Shen Y, Zikanov O (2016) Thermal convection in a liquid metal battery. Theor. Comput. Fluid Dyn. 30(4):275–294

Weber N, Galindo V, Stefani F, Weier T (2015) The Tayler instability at low magnetic Prandtl numbers: Between chiral symmetry breaking and helicity oscillations. New J. Phys. 17(11):113013

Weber N, Galindo V, Stefani F, Weier T, Wondrak T (2013) Numerical simulation of the Tayler instability in liquid metals. New J. Phys. 15:043034

Weber N, Landgraf S, Mushtaq K, Nimtz M, Personnettaz P, Weier T, Zhao J, Sadoway D (2019) Modeling discontinuous potential distributions using the finite volume method, and application to liquid metal batteries. Electrochim. Acta 318:857–864

Weber N, Nimtz M, Personnettaz P, Salas A, Weier T (2018) Electromagnetically driven convection suitable for mass transfer enhancement in liquid metal batteries. Appl. Therm. Eng. 143:293–301

Weber N, Nimtz M, Personnettaz P, Weier T, Sadoway D (2020) Numerical simulation of mass transfer enhancement in liquid metal batteries by means of electro-vortex flow. J. Power Sources Advances 1:100004

Weber N, Beckstein P, Galindo V, Starace M, Weier T (2018) Electro-vortex flow simulation using coupled meshes. Comput. Fluids 168:101–109

Weier T, Bund A, El-Mofid W, Horstmann GM, Lalau C-C, Landgraf S, Nimtz M, Starace M, Stefani F, Weber N (2017) Liquid metal batteries—materials selection and fluid dynamics. IOP Conf. Ser. Mater. Sci. Eng. 228:012013

# Correction to: Electrochemical Cell Calculations with OpenFOAM

**Steven Beale and Werner Lehnert**

## Correction to:
## S. Beale and W. Lehnert (eds.), *Electrochemical Cell Calculations with OpenFOAM*, Lecture Notes in Energy 42, https://doi.org/10.1007/978-3-030-92178-1

The original version of the book was inadvertently published without the ESM material. Correction to the previously published version has been updated.

The updated original version of these chapters can be found at
https://doi.org/10.1007/978-3-030-92178-1_1
https://doi.org/10.1007/978-3-030-92178-1_2
https://doi.org/10.1007/978-3-030-92178-1_5
https://doi.org/10.1007/978-3-030-92178-1_8
https://doi.org/10.1007/978-3-030-92178-1_9

© Springer Nature Switzerland AG 2022

S. Beale and W. Lehnert (eds.), *Electrochemical Cell Calculations with OpenFOAM*, Lecture Notes in Energy 42, https://doi.org/10.1007/978-3-030-92178-1_10

GPSR Compliance
The European Union's (EU) General Product Safety Regulation (GPSR) is a set of rules that requires consumer products to be safe and our obligations to ensure this.

If you have any concerns about our products, you can contact us on

ProductSafety@springernature.com

In case Publisher is established outside the EU, the EU authorized representative is:

Springer Nature Customer Service Center GmbH
Europaplatz 3
69115 Heidelberg, Germany

www.ingramcontent.com/pod-product-compliance
Ingram Content Group UK Ltd.
Pitfield, Milton Keynes, MK11 3LW, UK
UKHW021840270726
14058UKWH00002B/260

*9783030921804*